How Humans Became Intelligent

When did it start,
Why did it happen,
What else changed, and
Who are we now.

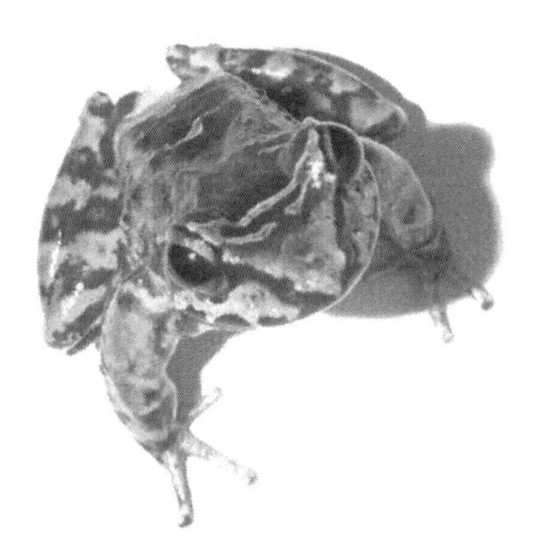

A CURRENT TECH PUBLICATION

How Humans Became Intelligent
Lynnette Hartwig

Requests for permission should be addressed to the publisher:
Current Tech Permissions Dept., 12018 Quito Rd. Richmond VA 23112-3771
or by email to Lyn8nn@gmail.com.

Library of Congress Cataloging in Publication Data
How Humans Became Intelligent / Hartwig, Lynnette
p. cm.
Includes bibliographical references and index
ISBN: 978-0-989178457
1. Evolution-History. 2. Human Evolution
I. Title
576.8 2017906878

Printed in the United States of America

20 19 18 17 16 15 14 13 12 11 10 9 8 7 6 5 4

This book is dedicated
to my son, Chapin, whom I love
most in the whole world,
and who is paying it forward.

Contents

Credits for photos on p. 209

The Paleolithic
↑ Pliocene (before *Homo*)
Lower Paleolithic (c. 3.3 Ma – 300 ka) Oldowan (2.6–1.7 Ma) Riwat (1.9–0.045 Ma) Madrasian Culture (1.5 Ma) Soanian (0.5–0.13 Ma) Acheulean (1.8–0.1 Ma) Clactonian (0.3–0.2 Ma) Middle Paleolithic (300–45 ka) Mousterian (150–40 ka) Micoquien (130–70 ka) Aterian (82 ka) Upper Paleolithic (50–10 ka) Baradostian (36 ka) Châtelperronian (45–40 ka) Aurignacian (43–26 ka) Gravettian (33–24 ka) Solutrean (22–17 ka) Epigravettian (20-10 ka) Magdalenian (17–12 ka) Hamburg (14–11 ka) Federmesser (14–13 ka) Ahrensburg (12–11 ka) Swiderian (11–8 ka)
↓ Mesolithic ↓ Stone Age

Ma means million years and ka means 1,000 years.
Do not study this chart. You won't need it.
Chart courtesy of https://en.wikipedia.org/wiki/Paleolithic

Foreword

I hate Forewords and Introductions in books. I've read less than 5% of them. Sometimes, if I particularly like the book when I'm done, I go back and read the Foreword. After that, I like the book a bit less.

They're like eating a big spoonful of butterscotch ice cream topping; I thought it would be tasty when I started, but now I'm just working through it and hoping it ends soon. I hope my foreword is different.

Have you ever driven on the freeway for a long time and believe you're almost there when the green sign looming overhead mentions a city, two miles ahead, that is 100 miles to the side of your destination? I have.

Have you ever hiked on a trail for many hours, then it slowly dawns on you that walking in this direction, as you have been for over an hour, is actually taking you farther away from where the trail you thought you were on should end? I have.

Once the thought dawns, suddenly memory produces a flood of other signs that should have registered, should have clued me to turn around way back then.

This is one of the truly crummy feelings in the world. Missed the turn. Missed eight other things that any half-thinking human being should have noticed an hour ago. Can't just shortcut sideways. Must go all the way back and then go forward in the right direction. No rest in the near future, only more work. The extra work is nothing compared to the shame of having messed up. If there are other people in the car or hiking party, it's either angry words or tense silence, right? You are mad at yourself as if the 'you' that made the mistake was a different person.

There are actually people who can't take it. They keep driving or walking in the wrong direction, refusing to backtrack. Their every fiber rebels: this HAS to be the way because I'm too experienced and too smart to have made such a colossal mistake. Not a tiny mistake with a ten minute loop-back, but half the day, with all kinds of other signs appearing every few minutes that should have alerted anyone. As the other people in the car or hiking party come to a consensus that this is the wrong way, Mr. Expert continues to persist.

Ironically, the oldest, most experienced and knowledgeable will be the last to change course.

It takes intelligence combined with bravery to turn 180° around and fix it without going a step farther in the wrong direction. Most people go a little bit longer, slow down, then freeze in their tracks for a minute or two 'to think.' Meaning to adapt to the new reality. The new reality being "I missed lots of signs, I'm not as observant as I thought I was, I didn't question myself at all, and now I can't hide. I have no excuse. My party knows it, and the people at the destination will see our late arrival and know it."

But guess what. We're human. Your friends will go hiking with you again. The family will get in the car and let you drive. It's crummy, like dropping the bleach bottle in the kitchen and the dog runs through the mess. That white paw print on the rug becomes a memento. Part of your life. Not wrecked, but more.

If you don't yell or behave badly, we'll all laugh about it in just a few months. If you do behave badly, then it will take a few years before the others forgive you and see the humor.

Your choice.

The Point

The point of a book is like the score of a ballgame. It's whatever one might say if asked what the book is about. Here are the conclusions up front. If you prefer to have the conclusions evolve like watching the whole game before finding out the score, then just skip to the next chapter.

Spoiler alert.

Evolution is of two types, forward-changing and back-reactivation. Forward-changing takes time, but back-reactivation is done in a handful of generations. Life today is the result of choices made by instinct, by 'thought' that is not thinking with a brain, but very powerful thinking all the same. This 'thinking' can perform complex statistical calculations in less time than our most powerful computers, and do it in a space not even half the size of a fruit fly's brain. Because this immense thinking capability fits into a tiny spot, nature prefers instinct over brains.

The environment does not demand adaptation. It does not cause genetic change. It is only the judge and jury after the fact, after a species tried something out. Therefore, the large majority of species variation we see today wasn't necessary for survival, it simply did not compromise survival very much. Nature loves the messy bird.

In nature, what's bad is good and what's good is bad. Nearly all life on Earth has incorporated the worst that happens, whether storms, fire, or bugs feeding on them, into good-for-me strategies. Most life on Earth is accustomed to several-hundred-year cycles. We see only a snapshot so define species by what we see. Most species will employ instinctive [and new to us] responses when other conditions arise.

Species in transition from one food source to another or from one environment to another can, if the change is challenging enough, develop a bit more intelligence to compensate for its previous instinct not working under the new conditions.

Changes related to brains and intelligence are done by 'more' or 'increase' commands applied to whatever brain the creature already has. Yet even after all these millions of years of life, less than 1% of life by weight has non-instinct thinking capability. It's not nature's first choice, and definitely not a goal.

Humans don't just use language, we all invent language. We enjoy altering language. How our throats work had to change to enable speech.

We are the water monkey. Everything about us indicates we left the trees to begin swimming. Our bodies, ears, hands, feet, and digestive system became adapted to deep-water hunting with spears and nets. We lost our hair because we swam. Our fingers became dexterous because we made rope, then nets, to fish. Like most land mammals that spend half their day swimming, our legs moved backward to align with the spine, making it easier for us to stand upright on land. Our feet became flippers. We used to have webbing between our fingers and toes, but very recently it was suppressed.

We had too many major life changes. First from trees to shore, then from vegetarian to fish-eating, then to droughts and ice ages. After the first bump in intelligence, we now had an 'increase intelligence' command in our recent past. It was activated one too many times.

We are nuts about weather. Fatherhood used to be more certain. Medicine has taken a funny bounce, one we must return to sensibleness. Men are too obedient for today's world.

See the Future. We must use our intelligence to stop the collision course we propel towards to our obvious fate.

Thinking Without a Brain

Water Lilies

Artfully, the leaves and flowers of a water lily lay atop the water, lazily floating without effort at the plane where water and air meet. To survive, the water lily must have standing water not too deep, not too shallow, not too fast-flowing, not too shady, not too cold, not too salty . . . the environment must meet all of their not-toos, but in the sliver of environment that remains, water lilies have bloomed and thrived for 30 million summers all around the world.

When water lilies grow, their height is not predetermined; nothing in their nature says to grow only six inches and then bloom, or another variety permitted to grow only eighteen inches and stop, or a third variety to sprout up six feet and then stop. Nature's code is more elegant than that. Genetically, each plant knows to keep its leaf in a tight coil while growing upward until it hits the light, the air. Once it feels that, the leaf is unfurled. It has no set distance to grow. Two neighboring lilies can float identically, but one is rooted in a fallen tree trunk and one in a depression so the length of their stems vary three feet in length despite being exactly the same age.

Stem length of a water lily tells you nothing about its age. The height of a pine tree does. A pine tree would no more decide to grow several feet in two months like a water lily than a water lily would decide to grow only nine inches this year like the pine. This seems very obvious and plain, but it is actually pure genius. What a plant or animal *cannot* do after ten million years of evolution is as much a genetic choice as what it does do.

After sending up a few of the energy-collecting leaves and ascertaining that it is safe, the lily sends up a flower shoot.

Pink lily patch.

The water lily thrives best with other same-species lily plants in the vicinity. The lily fertilizes its seeds by attracting flying creatures to its flower, the specially-created enticement the plant loads with sweet stuff for the flying creatures. It's even more attractive when the single trip provides several flowers in a tight group. A lily doesn't 'know' flying creatures are up there. The plant has no capacity to think about what to do if no flying things come, or work itself into a better location, or decide when to spread flowers and when to fold up.

The genetic code takes care of those things. A lily behaves in a very sensible, conservative manner to continue its life and reproduce, but not by thinking. It is all encoded, foreordained before the lily seed takes root. The plant's plan includes strategies for both the individual lily and lilies as a group, a lily patch.

Each water lily has built-in options for dealing with environmental issues before those events occur, even if they are happening for the first time in a dozen years. Every lily has triggers it can evoke if the water is shallow, is deep, or dries up near the end of summer, even if its particular patch hasn't seen any of those in six generations. Part of the normal genetic plan is to grow until it reaches sunlight, then spread its furled leaves. Once it reaches light, instead of being content with the straight vertical of its stem, it continues to grow a few more inches, causing the leaf to slide sideways so it does not rest exactly above the root ball. A smart plant knows it will need some safety slack. The water level of the lake can rise or fall by several inches yet the lily effortlessly retains a position atop the water. If levels rise more than that, the plant has some time to kick into high gear to grow more stem.

Every pond, lake, sea and ocean is a wide-mouthed bowl in the dirt. Like every wide-mouth bowl, each additional inch of water level height in the bowl contains almost twice the volume than an inch-high slice of water a foot lower.

Over millions of years and just as many instances, water lilies figured out the rule that we humans figure out with math, and this grasp of physics has been hard-wired into the plant's DNA. When the lake level rises, the plant is cognizant of the increase in verticalness of its stem and starts growing more stem. A water lily doesn't need the ability to grow two inches of stem in two hours; it knows each additional inch of lake water height comes at the price of a great deal more water than the previous inch. Growing an inch every six hours will serve the purpose.

In that nifty way of genetic-think, this is sublime; it prevents the plant from growing huge amounts of extra stem at every transient occurrence of higher water level, such as a huge rainstorm that causes lake levels to rise quickly but recede a few inches four hours later. Water lilies adapt to rising water levels only if they continue increasing over several hours.

White Lily patch, calming the ripples.

The enemies of water lilies are waves and choppy water, which can uproot them, since their leaves and flowers have evolved to be buoyant. In a mind-blowing devotion to self-interest—and so typical of nature that we think of it as obvious or inevitable—the water lily loves to hang with other water lilies, and the lily patch covers its corner of the world thoroughly with big wide leaves. Under this patch are dozens of flexible, somewhat-vertical stems. Both these vertical rods and the, let's call it a blanket on top of the water, serve to dampen waves and choppiness.

To add to the amazing cleverness of it, the plants on the perimeter obviously catch the full brunt of whatever surface action is going on in the lake, and sometimes they are uprooted and carried away. But that's a good thing, because unlike seeds, this is a hale and hearty full-grown plant that might get stuck on some bump along the way, reroot, and start a new colony.

The worst thing that could happen to an individual plant with the unlucky location on the perimeter of a lily patch is also a better way for the species to propagate by taking advantage of hurricanes, tornados, monsoons and other horrible weather to get to distant lakes and streams.

The water lily is proof that like much in nature, what's bad is good and often, what appears good is bad. Without insects that feed upon it, floods and hurricanes, the water lily would not spread as successfully as it does to new environments. It has taken all the bad in its world and converted it to good-for-me. Water lilies didn't get to choose how to pollinate or what kind of weather to have; they took the toughest circumstances that came their way and devised strategies to make those a positive. More importantly, the water lily did it without thinking. Meaning without having a brain and puzzling it out.

It happened via another kind of thinking. The kind of thinking nature prefers: we call it instinct, or genetic programming. It is a hard-wired response each lily 'knows' and will evoke, even if the circumstances are new to this particular plant. It's in the DNA. Plants don't need to 'think' to initiate quite complex survival strategies. They don't have to learn from older plants. It simply happens.

If the way water lilies group, reproduce, and calm the waters seem obvious to you and you can't imagine another way, consider if a rose bush adopted any strategy of a water lily, or a water lily of a rose bush; would it be helpful? If not, how does it know? Why did it pick this set of options and ignore hundreds of others? These plants have made many choices along the way. A water lily could do many things via 'genetic thinking' after 30 million years, including all the options of a pine tree or a rose bush, of seaweed and grass; it chose a few dozen key behaviors based on the information it had about its environment. It filled a niche it perceived was open–without thinking, without conscious planning–and in different lakes across the world expands into all the varieties that aren't a detriment to survival, such as displaying a range of color, amount of petals, and size of leaf, variation that doesn't amount to significant reproductive advantage. There appears to be variety simply because it doesn't reduce reproductive success, not because it improves it.

If genetic thinking is such a crude tool, a heavy hammer, a blind master, how do we explain the hundreds of varieties of water lily? How do we explain that in the same environment there may be flowers with a variety of colors and no specific color has a big advantage? Does instinct try different things just to see if they work? Does it try different things merely to keep the ability to change as a tool in the toolbox, in case it's needed one day?

Vital Spark. VS. This book will talk about living things from beginning to end. As a shorthand, living things will be defined as those which embody a Vital Spark. Not simply a vital spark within itself, but a vital spark that can be passed along, sent forward in time to entities that become clearly not the original organism anymore. This includes one-celled creatures, multi-celled creatures, cells that band together in brotherhood so all may survive, insects, plants, reptiles, animals and all things with features in two of those camps.[1] All entities with a Vital Spark shall be called VS in this book, to save a little ink.

Instinct, i.e. genetic thinking, works via a system that has no need to see the future because it knows the past. Not only the past several generations, but long into the past, back when the VS wasn't what it is today. Every single VS adapted, then adapted again, and then again. This isn't an educated guess. Sixty six million years ago Earth saw up to 75% of species die off in a cataclysmic event called the K-Pg extinction. This is when all the dinosaurs, except birds, died off.

---

[1] Classification systems for plants and animals that began in the 1700s and were firmed up in the 1800s were based on superficial physical characteristics that are increasingly strained as genetic information comes in. New information shows the branches of the tree of life did not connect the way the old timers thought they did. Overhaul of classifications is still on-going. Most of the changes involve recategorizing microscopic creatures, but occasionally even a mammal may be more closely or distantly related to another mammal than the old taxonomy charts alleged.

A huge percentage of the remaining 25% were one-celled VS. There are few VS of a size visible to the naked eye that have remained similar-looking and functioning since then. Changing just to try things out, possibly find a better niche, is the primary logical response after a cataclysmic event changes the environment drastically. All of today's species have that tool, change-just-to-check-it-out, in their genetic toolbox. The ones that didn't are in the 75% that didn't make it.

Species are capable of change for change's sake. Darwin had it a little wrong; he thought species changed in response to conditions. What we see, however, is that species change constantly and all the time. For 'just in case' the environment changes or there's a food source not being exploited.

In that light, sexual reproduction could be viewed as a means to pull mutating individuals back into the fold. It's a damper upon genetic drift, not a helper. Sexual reproduction is nature's solution to the tendency of every VS to become different than its distant ancestors in tandem with all the other VS. When a VS has, in rough estimation, 40 random mutations in 5000 generations to useful DNA (about 8% of mammal DNA is used, the rest is dormant), changing favorite foods, size and shape, a VS could not develop an instinctive knowledge of friend or foe, or of what's safe to eat.

Asexual reproduction is fine for one-celled creatures, but when multi-celled VS began to form, nature had to devise the most simple, fail-safe way to ensure a creature could know, by instinct, what its food supply was and recognize it when it found it. This meant it had to be in the best interests of most species most of the time to stay the same. Once they settle into a niche, they fill it for a million years or more.

Genetic variation is another matter. The longer a species has widely interbred within its type, the wider the genetic variation within the species. This is so profound and universal that scientists hang their hat on it. They have seen enough genes to create a chart of how fast in-species diversity happens. Using this chart, they can pin a date on when a sub-species was separated from the main group, or how long ago two similar but now not inclined-to-breed together-species diverged. In the near future we won't need fossils to do this; it will be done by examining the DNA of both VS. In fact, now that they know when a common ancestor was alive, they can look for it and identify it when they find it. In practice what will happen is a fossil remain found years ago and now sitting in a museum can be newly classified as the ancestor to today's species.

When a small group of a long-time stable species becomes separated physically from the main group, it changes. The food and weather can be identical. When birds can fly back and forth to an island several miles off the coast, that island will not create a new sub-species regardless of the size of the island. If that same island is 100 miles away, in time it absolutely will.

This smaller genetic pool means some recessive genes have a better chance of prevailing. It means a trend that might have been developing in the main population becomes accelerated. While there might be a large amount of variation in the main population in say, spotted fur or feathers, by random chance half the individuals on the island may be prone to larger spots. In 500 generations they might evolve to have only large spots, or even a solid color. Not because the environment needs that, but because the only difference it made to reproduction was in the eye of the beholder, i.e. the opposite sex. Zebras became striped not because their environment demanded it–there are several other hooved animals in the same environment that are not–but because it did no harm and the ladies might have liked the stripes.

All VS have genetic variation before needing to adapt to anything. The change endures when the environment tolerates it. The environment is never the cause of change; it is the judge and jury after the fact. The environment does not whisper or hint 'more pointy beak!' or 'pinker flower!' or 'bigger claws!' The creature itself must make the change for change's sake, and after that, if it lives, it gets to reproduce. If it simply makes no difference, it still gets to reproduce. If it does something outlandish like develop yellow feet (duck) or a wattle (turkey) or goes bald (hippopotamus), but none of that harms reproduction, it gets to continue.

All of today's species have genetic matter with thousands of turned-off genes. Despite the appearance of permanence and a fossil record going back millions of years, most species have it in them to change in profound ways at the drop of a hat.

The Nature of Nature

"Propel, propel, propel your craft softly down liquid solution. Ecstatically, ecstatically, ecstatically, ecstatically, existence is simply illusion."

-- Fred Rogers, American television host and producer, (1928-2003)

The title of this chapter is the epitome of the problem faced in writing this book: the difficulty of selecting words that convey the same meaning to everyone when English words share so many meanings. It's not 'simply illusion.' All of the data and most of the thoughts in this book have been revealed in publications before. They were buried in details about the process or worded so erudite that it slipped by most readers. Like Mr. Rogers, I can write eruditely if I have a mind to. But I won't.

In the title above, the first 'nature' means tendency or inclination. Perhaps it implies 'the type' or 'the temperament.' The second 'nature' means wildlife and undomesticated plants. Many people include only plant life left to its own devices; nature is not a mowed lawn. Other people encompass insect and animal life along with plants. Still, nature is not a dog catching a Frisbee. Nature is one of those words whose meaning is so much more than the dictionary states.

To make it even more thorny, we are inclined to use the word 'nature' for only land environments. We don't conjure up visions of sea-bottom life when we hear the word, but if asked we won't deny that there's nature in the water too.

Nature is a friendly word; we leave the city to 'visit nature.' My mom once complimented my house by saying 'there's so much nature here.' I knew what she meant.

Nature has dark side, though. When a hawk swoops down on a sparrow, we shrug, "That's nature." When a cheetah takes down a wildebeest, that's nature. When an ice storm fells some trees, that's nature. When termites burrow into the wood corner beam of our house, that's nature–no wait, forget nature, that's a phone call to the exterminator. When our car gets a flat on the freeway ... that's not nature. That has nothing to do with nature.

I could have great fun with all the things that aren't nature, but the point is made: nature is what life on Earth does without human intervention, in efforts to cope with the environment and its niche in the world. It is a force of give and take, of wax and wane, of win and lose, of nice days and awful days. Nature is a unity beyond human comprehension. Nature is a tree making shade for other living things when its intent is not kindly. In following its greedy, light-hogging strategy it creates a hospitable environment for many plants and animals that cannot survive without some shade in their lives, a home for creatures that can't thrive on the ground, and some place for little birds to nest.

In Nature, the good is bad and the bad is good.

Example: One cold winter day I watched a bird eat the birdseed I'd poured in a row atop the rail on my deck. The seed was clearly in sight and no competition was around, yet the bird ate with quick, jerky dives of the beak. This caused some seeds close to the aimed-at seed to skitter away, flying off the rail. Silently I chided, "Little bird, if you ate in a neat and calm manner, you would be able to eat every seed. You would have better odds of surviving the winter."

Then it dawned on me. Nature doesn't want any neat birds; nature wants messy birds. A neat bird does its offspring no favors. Every seed flying sideways, whether on the rail or still connected to a bush, has a chance of becoming a new plant that provides hundreds of times more food in a few years than a single seed does today. Neat birds that ate every single seed instead of only 85% of them would be extinct in twenty generations. Nature loves the messy bird.

Take the case of trees: if big trees grow unchecked by storms or fire, in time they cover every inch of the sky. The plant and animal life below them diminishes to only those able to thrive in full shade. No new trees can get a foothold on life without sun. In temperate climates the area becomes a barren place under the trees, at least until an ice storm or fire takes them down and allows a fresh start. Still, we consider trees 'good' and those natural wildfires to be 'bad.'

Are forest fires bad for trees? Are they bad for the wildlife living in the woods? Around the 1920s when hardware and technology gave Americans the wherewithal to squelch forest fires for the first time in human history, we started doing that. We were saving the forest. We were saving the critters that lived in the forest. We were certain this was the truth.

What happens to a forest in the US if there are no fires for 120 years? That question went unexplored. The assumption was 'it will be nice, just like now.'

Turns out, it's the opposite of nice. Trees grow taller and wider, blotting out the light. The diversity of groundcover diminishes. The forest becomes mainly rotting-leaf blanketed dirt with a fern here and there. Creature life diversity also reduces; as food supply disappears there are fewer species of butterflies, birds and furry animals. Trees die and rot where they stand. Animal life is dominated by rodents, and plant life by moss and lichen. New trees cannot sprout. Berry bushes, wild grapes, flowers? Forget it. Harvesting the tallest trees may have saved the forest twenty years prior, but the lack of sun and preponderance of moldy logs has now made logging the remaining trees untenable, both for accessibility difficulties and the lower quality wood in the still-living trees.

The only thing that will fix this woods is a good fire. Hard to believe, but true; fires are a necessary part of the cycle.

An example of this is Yellowstone National Park's 1988 forest fire. While thousands of people made efforts to contain it and keep it away from inhabited areas, in the end it was only rainstorms and then snow that extinguished it. A common tree in Yellowstone, the slow-growing, long-lived lodgepole pine, has seeds that are only opened by high heat.

No fire, no new lodgepole pines. There are species of trees that shoot up fast and provide some noontime shade in three or four years but die in forty. In the comfort of that partial shade the slow-growing oaks, elms, maples, lodgepole and walnuts get their start.

A woods isn't a static thing; it is a cycle. We see only part of the cycle during our lifetime. The existence of oil and coal indicates this cycle has been going on for millions of years. It's easy to imagine the first fronds or shrubs emerging after a primordial forest fire getting stepped on by passing wildlife, pushing them into the inches of muddy carbon dust, their leafy shape preserved for millions of years until we dig up that coal.

Fossil fern in coal found in Morocco.
Image from Wikipedia Commons.

The Yellowstone experience gave us another insight. Due to so many years of preventing forest fires, that fire was scary in its ferocity and thoroughness. A forest fire occurring forty years after the last one has less dead matter lying about for fuel. Green leafy life with good water content resists burning, so when those exist there are splashes and pockets of surviving greenery days after the fire. In the normal course of things, some trees in their prime will soldier on after the flames pass. Forests long beyond their time for a fire are not so lucky. Every single tall tree is consumed.

We had it all backwards. Our goal was to set aside nature enclaves full of diverse plant life feeding a variety of bird and animal life. By preventing routine forest fires we destroyed that very habitat for hundreds of species.

Near my home in Massachusetts[2] there is a nature path through the woods; in 2004, near the remnants of a stone foundation that was a home until the early 1900s there was still a lilac bush, herb plants and peony bushes where the garden had been. By 2015 they were gone, wiped out by lack of sun and probably lack of bees no longer venturing that far into the barrenness anymore. The path itself became impassable; town budget cuts eliminated the fellows with chainsaws clearing the huge fallen tree trunks blocking the way. It isn't a nice walk anymore; the plant life is mainly ferns and moss punctuated by tree trunks with no branches below twenty feet. It's spooky. All the wildlife action is happening along the edges of the mowed areas of my subdivision. That's where the birds catch insects, the berry bushes grow, the bunnies hop, and the deer graze near sunset. This 'nature' preserve wouldn't support wildlife at all if we didn't mow a strip twenty feet deep from the road!

The tale about forest fires makes several points about nature: one, human beliefs can be based on a snapshot of a long cycle that we don't understand. Two, we persist in discounting the evidence of our own eyes for a very long time. In the case of 'nature preserves,' the very thing we believe we are preserving, we are destroying. Perhaps the woods are too close to homes to start fires today, but it may be possible to log the bigger trees, thin it out to provide more habitat over

---

[2] I wrote this book over the course of several years, moving from MA to VA in that time. I've left references to MA as they are because it's the observation, not where I live now, that's relevant.

the next fifty years. Sorry for the digression; I can't help but ponder a workable solution even as I describe the problem.

We categorize lightning, ice storms, and hurricanes as interference with nature, damaging to nature. It is hard to wrap our minds around the fact that life on Earth is so accustomed to 'what happens' on Earth that it has incorporated the 'bad stuff' into being good for species even if bad for specific individual members of a species.

Life on Earth has adapted not only to seasonal variations, but to circumstances that an individual member of a species may never see in their lifetime. Because it sounds so farfetched–for example, say I just blurted out that forest fires are great for birds–the gut reaction is to snort. Yet it is easily proved by empirical evidence: have there always been lightning fires? Yes. Is there any evidence that a particular forest fire caused the end of a species? No. None. What does a forest look like twenty years after a fire? Great! Teeming with life. Noisy with birds calling out. What does a 100-year-old no-fire woods look like? Silent and moldy. That's empirical enough.

In fact, while reading several articles on the 1988 Yellowstone fire in research for this reference, what began to gel was how little the forest was bothered by the fire.

Observers commented that the roots of most plants were just fine and shot up new shoots in days. Fewer than 1% of soils were heated enough to burn below-ground roots and seeds. The charred ash became a wonderful fertilizer. Aspens and other trees that had been squeezed out by the prevalent lodgepole pines took root and flourished. Elk love eating young Aspens, so they are faring better than they were before the fire. Thank goodness the fire saved the park.

This illustrates that if nature had 150 or 200 year cycles it could mean what we know about a species isn't the whole story. A species could well change its tune–develop a tolerance for a new food, switch from diurnal to nocturnal, go from sleeping on a branch to sleeping in a tree crevice–simply due to where their forest was in the cycle. As a habitat goes through cycles, it's conceivable that some species would eat things and do things they wouldn't do fifty years ago when the forest was in another stage.

Yellowstone, a year after the 1988 fire. Wildflowers abound. Plants regrew in days after smoldering stopped. Due to the quick rebound, no reseeding was done. The National Park Service and the American Institute of Biological Sciences determined that less than 1% of the large wildlife population–elk, mule deer, moose, bears and bison–died as a result of 36% of Yellowstone burning over two months.

While our altruistic intentions are to save wildlife habitat, we are not changing our behavior when we see species disappear. It's a story repeated in hundreds of 'nature preserves' across the US. Creatures missing from quiet, moldy woods are dead from lack of food.

Life: what is it? An old oak tree will crash down in a storm, and several of its branches digging into the ground will sprout new branches pointing straight up. Did the first tree die, or merely change in mass? If the 'center' of life moves from a larger area to a smaller one, is that a continuance of the original organism, or can it be declared definitively that the old tree died and these are brand new trees? If a bacterium splits into two via binary fission and instantly after the split one is eaten, can the original bacterium be said to still be alive, albeit in only one of the halves? From the point of view of 'life' there is no death until every last one dies. There is only handing off from bigger to smaller, from an older body to a new body. Even if there are huge species die-offs, as have happened in the distant past, whatever survives–doesn't matter what–will begin changing to fill the ecological niches now emptied.
Leaf-eaters will become predators, swimmers will walk on land, land creatures will return to the sea, short plants will become tall plants, burrowing critters will climb trees. In ten million years the mix will be different but fill the same niches with the same behaviors as the died-off species. The teeth, the feet, and even the internal organs will slowly become a match to a VS that held that niche in that climate at some other time or on some other continent. The heads of a deer and a kangaroo are almost identical because they eat the same type of food. The jarring difference in how they chose to outrace their predators emphasizes how originating from two very different ancestors could lead to identical creatures, but not always. The features unrelated to eating could remain as they were, or adapt to subtle differences in terrain in a genetically easier way.

When a niche is available, nature doesn't know or care what's done on other continents. The niche filled by birds in one place is filled by bats in another, by marsupials in one with mammals in another, by insects in one with hummingbirds in another. The relevant body parts' color, size and shape changes to fit the task, becoming stunningly similar. Only when the other parts still vary do we find it a novelty. The niche filled by ducks on other continents is filled by the platypus in Australia. It has a bill that looks and functions like a mallard duck's because its food and method of securing it, by mouthing the mud of lakes and streams, is almost identical to that of some ducks.

The majority of Australia's marsupials are so identical to placental mammals that it was long assumed the pouches alone were the adaptation to that continent. The reality is that marsupials and mammals diverged over 125 million years ago. In the process of filling the same niches, those adapting animals developed the same feet, teeth, hair, ears, size and shape as their other-hemisphere counterparts.

How can the identical survival waltz be created from scratch by different means on different instruments on different continents? The enormity of it seems impossible. Yet viewed another way, creating an entirely different waltz is what would be improbable.

Say I'm beginning life on this planet from scratch. To not bore you with starting from single-celled organisms, let's fast forward until there is grass. Lots of grass. Naturally, some VS will develop that eats grass. In response, some grass will make itself untasty or too hard for that creature to eat. Or raises its green parts too high for the grass-eating creature to reach. The area still needs grass-eaters, but some climbers and some tall big-mouthed creatures that eat the stems will develop.

Another creature will develop to eat the critter that eats the grass. Insects live in the grass. A creature will develop to eat the insects and their eggs. Insects will mutate to foil those eaters, and the eaters will mutate to overcome those impediments. Worms live in the ground. Some creature develops to eat worms.

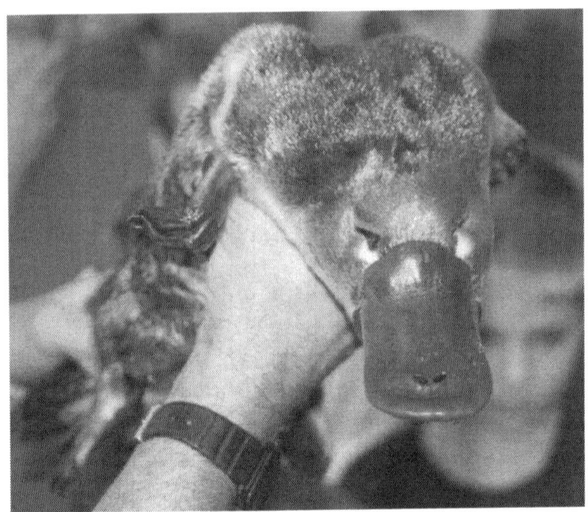

Duckbilled Platypus. As a swimming, shore-living creature it shares fur and feet with the otter and beaver; exactly like the duck, the bill is used to work the mud near the shores of lakes and streams for insects, eggs, and hiding small water creatures.

The Australian marsupial Thylacine, a.k.a. Tasmanian Tiger. It and African dingos look similar because they fill the same niche. It was called 'tiger' because it had striped fur. This marsupial is much more closely related to the mole in the next photo than to any placental wolf or dog.

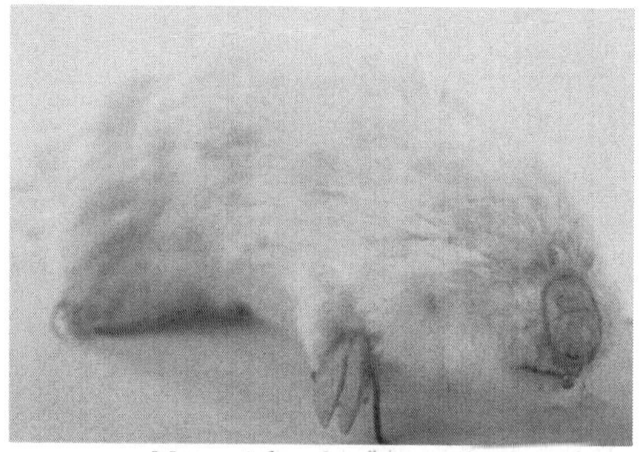

Mursupial mole of Australia

Golden mole of South Africa (mammal)

Marsupial mole physiology is stunningly similar to that of mammal moles, even though their ancestors diverged longer ago than ours diverged from the giraffe and the house cat.

Coast Mole (mammal) of the US (Pacific)

Granted, the small story above is missing a lot of steps. But whether long or short, you get the idea.

When equilibriums are reached between the size, the appetite, and the reproductive quantity of species on that patch of land, that optimal configuration can last for millions of years. If a hunting species is too successful and breeds too much, the prey dies out and so does the hunting species in short order. Both niches will be expanded into by neighboring species over time. The species that pushes into the vacated niche may be closely related to the extinct one, or may jump from mammal to bird or lizard to mammal.

The earliest evidence of life on Earth, fossil bacteria that clumped in mats near thermal vents in the ocean, have been tentatively dated to 4.22 billion years ago, which is only 340 million years after the formation of Earth itself. It's estimated to have taken over 900 million years for one-celled creatures to band together to form multi-celled VS.

Since then, all over the world, when there is the same kind of food, lighting and gravity, genetics will optimize the relevant body parts to very similar solutions.

A garden or woods has hundreds of flora and fauna, from microscopic to a hundred feet tall, each playing their own game, and most of them have been filling their niche for three million years. There is nothing unsuccessful about staying in a niche for three million generations, like the water lily, each generation employing the same life strategy as their ancestors did with the same inevitable success.

Full mastery of life is to survive when volcanos, floods, hurricanes, ice ages and fires take their toll.

Intelligence is not a requirement or even an obvious advantage to survival. Life forms that move around are a dinky percentage of total life by volume. To be generous, let's say VS that propel themselves by legs, fins, wings or creeping along make up 1% of all life as measured by the pound on land and water. Moving-around life tends to have a nerve center, a control central, sometimes complex enough to be called a brain. Pound for pound, living things without brains or nervous systems are 99 times more predominant than those with them.

Now sort out, pound for pound, the moving-around creatures that operate mainly on instinct. These are the ones that need no training from a parent. A parent might guard them and feed them for a short while, but there is no training in what to eat or not eat, what is a predator and what is not, what to do in the winter, what to do if it rains, and where to sleep at night. They get some protection or are simply birthed or eggs laid, then abandoned. Insects, crustaceans, worms, fish, reptiles, birds and amphibians are all mainly instinctual and make up over 90% of the volume. That leaves brainpower as a survival tool employed less than one ninth of one percent of the time.

If having a brain was a guaranteed-success sort of feature, it would be popular. What a VS does not do after 30 million years is as significant as what it does do. If moving around, with all its attendant risks, decision-making and conflicts were the big goal of life, that doesn't explain why only 1% of life on Earth went that way.

My approach to the digestion of facts from anthropology, archaeology, paleontology, biology, nutrition, genetics and psychology is to bounce them off each other for discrepancies and mismatches. Some are glaring. Some are absolutely irreconcilable between the fields. A theory needs to fit all of them, not just most of them. The greatest clarity is achieved using a rule made famous by Judith S. Sheindlin, a.k.a. Judge Judy, from the long-running TV show. What works when listening to testimony of partisan litigants also works when holding up statements from one professional field to those of another professional field. I call it the Judge Judy Rule of Theories: If it doesn't make sense, it isn't true. If it doesn't fit with all the facts, it isn't true.

Perhaps some information in this book isn't as current as it could be, and some descriptions may not include all the nuances of the longer story, but hold me to the Judge Judy rule of theories instead of proceeding with quibbles about ages or dates or names or places. The problem with a theory that no one believed prior to hearing it is that facts which were imperfectly shoehorned into other theories must now be extracted and placed, better-fitting, into the new one. This is an action that doesn't happen overnight. It might take a year. Or two. In which case you will feel better about yourself if you do not plant your heels so hard against it upon first hearing, but allow yourself time to ponder.

Eyeblink

Peppered Moth

The story of the peppered moth of England provides a fascinating insight into how quickly evolution can move. In 1848, the peppered moth had a camouflage coloring of light grey with little speckled dots, to blend in with light-colored lichens and tree bark. While each one had a different mix of grey and speckles, peppered moths were mostly grey. Then came the Industrial Revolution, and the land between London and Manchester, over 200 miles apart, was blanketed with soot from the new coal-burning factories. Not only did the soot kill off much of the light-bodied lichens, it coated the tree trunks, where the moth was accustomed to hiding, with soot.

Typica and *carbonaria* morphs on the same tree. The light-colored *typica* (below the bark's scar) is nearly invisible on this pollution-free tree, camouflaging it from predators.

Biston betularia f. typica, the white-bodied peppered moth.

Biston betularia f. carbonaria, the black-bodied peppered moth.

Peppered moths today

By 1895, less than 50 years later, 95% of the peppered moths in this area were black with white speckles. Some estimates put the frequency of black-colored moths prior to the Industrial Revolution at about .01%. What we know today is that black coloring is a dominant allele (half a gene) and light gray is a recessive allele in peppered moths in this part of England. Even though things are not so sooty today, there are still plenty of black ones around, thanks to the dominant gene.

The prevailing theory today is that black *was always* the dominant allele, even when the entire reproducing population was grey. Before the Industrial Revolution, black sure behaved like an extremely rare and recessive gene. If you believe the 'black was always dominant' theory, then you believe that for 100,000 years, peppered moths are popping out both black and grey, but the black ones are being eaten by birds without reproducing, and before any human spots them, so 99.99% of the reproducing adults are grey. Only .01% are black. Then, in a span of less than 50 years, the exact same bird-eating stimuli finally works to change the species color. Whew. How much longer could this species endure without Industrial Revolution soot?

I'm invoking the Judge Judy rule. A species is not likely to keep an always-eaten-up trait dominant for 100 generations, much less 100,000.

Not only did the peppered moth get a color overhaul, but in a span of less than 50 years a recessive gene became dominant. This makes sense. It has to be the only way any species diverges from its previous population and acquires a characteristic that is adaptive to a new niche.

There's no doubt this happens because 97 million years ago my pet dog and I had the same grandparents, our lines diverged, and today our dominant traits are very different. Recessive genes becoming dominant genes are the means to accomplish changes that stick. I submit that the amazing thing that happened to the speckled moth is *not* the color change, but rather that the recessive allele became a dominant one in the species *in less than forty generations.*

The peppered moth is a profound example for another reason: there is no genetic evidence that all of today's dominant-black allele moths are descended from one individual experiencing a genetic mutation. For this to have happened means an epicenter of black moths formed, were successful, and spread out from there. The children of that first dominant-gene black-winged female would flutter off unmolested by birds to reproduce.

Walking that one-mutation theory out, let's say another factor that changed was the distance a peppered moth traveled before laying its eggs. Instead of laying eggs within a half mile of where she was born, as has been typical all along, let's assume the first generation females traveled up to three miles in a radius around their birthplace. Then assume each of those young also traveled up to three miles from their birthplace to lay their eggs. In this manner it would take 40 years to reach 120 miles from the epicenter.

What happened did not look like that at all. The moths turned black all over in a non-radiating fashion in an area hundreds of miles wide in less than 40 years. How far they travel from their birthplace to lay eggs didn't change. There is no evidence that today's moths are descended from a single mother only 150 generations ago. A 'sole mother' population has a tighter range of gene variables that would be detectable by those working in genetics.

The genetic footprint of moths with a dominant black allele retains the diversity of a normal population. Eyewitness accounts support that the transition was related to the sootiness of each industrial hub. It was as if a specific environmental change experienced by several populations of a species caused the same genetic adaptation to spontaneously occur.

Put another way, the species had within it a predisposition to handle environmental issues in a certain fashion, i.e. have darker-colored wings. Life on Earth is very complicated, with its DNA and RNA and resident bacteria.[3]

Every year we discover more complexity and depth in what is actually happening at a cellular level in all living beings. In our DNA there is quite a bit that today's scientists call 'leftover from before' or 'doesn't do anything.' That's like

---

[3] Bacteria that are vital for the creature's survival but are not reproduced genetically by the creature, but passed down from the mother.

standing in a long hallway with doors lining either side and saying only those I've opened contain a room. The others don't go anywhere. In time we will open all the doors to see what has been closed for now in our species, and in many species.

Gene research has found there are a relatively small amount of genes in the genetic makeup of VS. If you read books and articles that are fifteen or more years old, they will talk about millions of genes. At one time not too long ago it was estimated that it would take more than a dozen years to document the full human genome! The tools improved, and more significantly computers improved. Once the process became automated, it really zipped along.

Today we know that mammals have about 20,000 genes. We have documented the full genome of not only humans but dozens of species. You can even send a cheek swab in the mail and pay a little money to get a DNA analysis of your ancestry. If you do, you'll notice they still aren't good at guessing the color of your eyes or your shoe size. That information is in there, but it's a bit more complicated than simply spotting a particular gene.

While we may have 20,000 genes, we have millions of regulatory controls for those genes. A certain gene may encode making bone, for instance, but it's the regulatory material that dictates how long, how thick and when to start and stop growing. Genes are like the ingredients, and the regulatory controls are the recipe.

So a person could have genes for blue and green eyes, but one would have to access the regulatory information to see what's turned on or off. Right now plenty of geneticists are doing that–but not with eye color. They are pursuing diseases and medical conditions to save lives. So don't be too disappointed about getting the eye color wrong.

It's a mouthful to say genes and their regulatory controls, and most of us were raised thinking genes held all our information. So when the word genes [a distinct segment on a chromosome], or alleles [half a gene, one from each parent] or chromosomes [the snaky things, of which humans have 23 pairs] are used in this book, the meaning is identical to that from fifteen years ago: the genes controlling a feature or aspect of a VS.

Somewhere in the as-of-yet-not-pinpointed genetic code there are directions for a recessive gene to become dominant. There are switches for turning on previously turned-off genes to restore features that were once useful. It is plausible there is code for several different means of handling a variety of environmental changes. Therefore when an environmental change happens–gets hotter, gets colder, plants change color–a whole species can 'mutate' in unison. There will not be an epicenter, a single individual from which the change radiates.

It's as if it 'occurs' to dozens of impacted individuals to genetically change in exactly the same way. They can do this because they're not doing something new; everyone is reaching back, far back, in its genetic history for a strategy it can fish out and re-employ. This is what the Peppered Moth did. It converted to a new color in a proverbial eyeblink genetically while retaining the full genetic diversity of the old species. Each individual is fully able to interbreed with the others regardless of color. In fact, if one wanted to waste horrendous amounts of money on moth genetic studies, they may find two individual black moths, say one in Reading and one in Liverpool[4], that do not share a common ancestor for 300 generations, while a black and a grey moth in Liverpool may be only five generations separated.

---

[4] About 200 miles apart.

Permutation & Perfection

Turkeys

If you are under 40 years old, you will find it hard to believe what I'm going to tell you about turkeys. You will think that everyone had it wrong back then because it's different now. I assure you, what they said back then was right, as is what you see today with your own eyes is right. It's the turkeys that have changed.

The wild turkeys in America were loners. Because of their well-known habit of fleeing at the approach of humans, the Apache Indians considered them cowardly. They would not wear turkey feathers or eat them for fear of contracting the spirit of cowardice. Turkeys across North America were like this. Even when there were sub-species by geographic area, being shy of humans was identical.

Domestic turkeys, the ones we buy in stores, are descended from the Central American turkeys the Aztecs domesticated in the Tropics. Early Spanish explorers took several of those turkeys home. Over the years they evolved into bigger versions in Europe, and later on were shipped to North America as a barnyard bird along with chickens and ducks. The wild and domestic turkeys in the US vary more than merely 1000 years of domestication; their ancestors were from a different climate and possibly hundreds of thousands of years apart. Attempts to release domestic turkeys to the wild in the US always failed, not only because they had been domesticated too long, but because even if they reached deep inside for their inner wild bird, they had been a tropical bird so none of that instinct applied in snowy North America.

Wild turkeys in the US were considered an endangered species until the 1980s. By the early 1900s only 30,000 existed.

Once abundant across the US, as settlers cut down the forests, their habitat shrunk. By the 1950s, eighteen states had not a single wild turkey, not counting Alaska and Hawaii which never had any turkeys to begin with.

School children were taught that wild turkeys will not reproduce if people are nearby. They require wooded seclusion to survive. A couple of blocks of woodsy area won't suffice; it must be miles and miles of isolated territory. Because they are loners,[5] they are a fragile species when their numbers thin out. The odds of finding a mate during breeding season significantly diminishes.

When researching this section of my book, I found many sites credited Fish and Wildlife State, Federal and Activist organizations for their efforts to understand turkeys, making their return successful. What really made a difference, in my opinion, was creating a good mechanical device for capturing wild turkeys without seriously harming them, called the 'rocket cannon net.' Now people could capture small groups of turkeys and truck them to a new place together. The turkey habitat research? Kind of pointless since they lived wherever there were trees. Surviving their contact with humans unharmed, as it turned out, played a huge role.

One website profusely thanked a foundation for its extensive work in studying the habitat needs of turkeys, attributing the success of the redistribution project to their seminal work. I find this funny. I am laughing. There are seven million wild turkeys today. Just what part of 'habitat study' made that possible?

I live in a woodsy area where a pair of wild turkeys were tossed off the back of a truck in 2004. Or set down nicely and told 'shoo shoo, go on,' if you prefer. Nothing was done to make the habitat right before they arrived. It was just a couple

---

[5] It's probably the case that the loner theory was weak because even the native American's description said they fled before anyone could get a good look at what they were doing.

blocks of woods in Massachusetts with a creek running through it. It's a woods that started growing about the same time as most of the woods in New England: in the late 1800s after the farmer was too old to farm. All his kids got jobs in the Industrial Revolution or moved West, so it went fallow.

Rocket Cannon net trapping turkeys for relocation.

I wish I could say someone from the DFW, Massachusetts Division of Fish and Wildlife, popped out once or twice over the first few weeks to fling some corn onto a lawn to supplement the newly-arrived turkey's diet until they learned the lay of the land, but that didn't happen. Over the next years they tossed out a few more pairs of turkeys.

This isn't an isolated woodsy area; we have street signage that calls us 'thickly settled.' Houses and yards surround the entire woodsy area. A few houses, like mine, are in the middle of it. The woods itself is cut into four sections by roads and large grassy areas.

Did the turkeys suffer stress when we had lawn parties with loud music? Did they refuse to lay eggs because of the lawn mowing and kids playing? Nope. By 2010 we had a 'turkey problem.' Herds of turkeys mill on the road, stopping traffic. They can take over a front yard, pinning the homeowner inside. Scare them away with a loud noise? Car horns ten feet away will turn their heads and prompt slow movement at best.

Herd of turkeys not caring while people walk by in Massachusetts, fall 2010.

Pardon me, I'm laughing again at the thought that someone got paid for studying the habitat needs of a turkey. As far as I can tell, their only habitat need is food. Not only does their breeding seem unaffected by living near people, they actually do not even have an aversion to us. They are perfectly OK with sharing my driveway with me, as long as I stay on my side.

I'm not really complaining about the turkeys. I love that the species has rebounded. I credit their rebound with the rebound in fisher cats and foxes too. The 'turkey problem' may self-correct, due to the healthy reappearance of those predator native species. As much as some humans would like to credit other humans who 'studied' and wrote papers on turkey habitat, studies didn't change these turkeys.

What happened to wild turkeys is that a change occurred inside the turkey head. There is no evidence it started with one turkey, then spread to that turkey's children, then to those children, until they all had it. All the human-tolerant turkeys aren't related. It didn't just happen to, say, Indiana turkeys and work out from there; it happened to all the wild turkeys over the span of merely 30 years.

Wild turkeys went from being a scaredy-cat loner bird to this herd animal that goes wherever it wants. Not just in Indiana, but in Massachusetts, in California and everywhere. It went from a species that could be thrown off its reproductive game by a train horn in the distance to one that congregates with five half-grown chicks near a human's mailbox and moves along as slow as molasses when I display my very threatening human hand gestures and loud voice.

What the heck happened? How did turkeys all over the continental US change so dramatically in 30 years, from being endangered with extinction due to habitat disappearance to wandering about our backyards even when we play the radio on the deck?

When a habitat of the past million years goes poof in twenty turkey generations, current thinking is that evolution can't help. It's too short. But evolution can help. Evolution can reach back. It can cause the same change to evince itself in dozens of populations that have not interbred for 100 generations but are facing the same habitat-loss situation.

In the case of turkeys, a switch was flipped. Humans were recategorized from 'instinctively dangerous predator' to 'not a predator, just a noisy nuisance, like a squirrel.'

We didn't deserve that demotion. We are predators. Heck, we EAT turkeys. They should still be scared of us; we're still bad, even worse than the Apache who wouldn't eat them or use their feathers. How can they be afraid of the Apache but not of us? It's not like we don't harm them; we hit them with cars; kids shoot at them with BB guns–oh yes they do! It was prudent to be afraid of the Apache because they came around only a couple of times in the turkey's lifetime. Today, humans are here all day, every day. Instinct decided the cost to benefit ratio had disappeared. The instinctive fear was no longer helping the species.

One bit of code altered, and the species is saved from the brink of extinction. Elegant, simple, and hits the spot. It was possible because turkeys have the genetic ability to add or subtract a predator from their instinct. It is not implausible that in the past, new predators arrived in an environment and turkeys needed to add them to the instinctual list of flee-from critters. In fact, the arrival of humans to North America 15,000 to 25,000 years ago was the cause of their addition. Removing a predator from the list is probably done far less often. Due to the way the genetic switch works, or the fact that humans had turned it on less than 25,000 years before, it was possible for turkeys to simply flip it back to where it was 26,000 years ago.

Humans, meh.

Geese

The story of wild geese is parallel to that of the wild turkey. Up until the 1990s, Wisconsin's DNR allowed only a limited goose hunting season and issued goose stamps by lottery so the species would not go extinct. Today, well I don't have to tell you how it is today in the US. If you've walked in a park or gone golfing recently you probably still have goose poop on your shoes.

Geese are not great at reproduction. They start with six to nine young ones and are lucky if two make it to reproductive age. In their lifetime they're doing pretty well if they have six surviving offspring. It's simply impossible that all of today's friendly wild geese from South Dakota to Massachusetts are descended from one goose born less than sixty years ago. In fact, it's unlikely that all the friendly wild geese alive today had a common ancestor even a thousand years ago.

The thing they share is that their environment changed in exactly the same way: humans everywhere. All the individual geese had the same 'predator' switch they had turned on when humans appeared in North America. When the fear worked against their survival, nature turned it off. We still kill them. They simply don't react as if we are a predator. They mosey off as if we are a deer or a turkey when we approach, slowly but with the evil eye as if they have half a mind to peck us just for being annoying.

At times it appears as if all their 'friend or foe' decisions have gone from instinct to thinking. For instance, they are not afraid when they see my dog on a leash, moving out of the way only enough to be twenty feet out of reach. But when the dog is not on a leash they fly off as soon as they spot her a half block away. They evaluate her as a predator based on speed of approach, by thinking about it, not simply reacting instinctively at the sight of a predator.

Global, identical genetic change can occur in a species rather swiftly when 1) there is genetic precedent of a very similar sort within a few million years, and 2) when the environmental change is happening to thousands of members of the species. This genetic or instinctual change will occur in populations experiencing the environmental issue. If there were pockets of the species that did not, such as peppered moths far from industrial centers or turkeys living in truly isolated forests, it stands to reason they wouldn't change. Yet both populations would not show a reduction in genetic variation indicative of a reproductive bottleneck, i.e., a sole mother. The genetic change would not affect the ability to interbreed because both types had all those genes before and now.

Why is this an important theory? Because scientists like to stick a pin on the map where genetic variation started. What makes more sense is marking the location with a wide paintbrush on a small map.

The sole-parent situation accounts for a percentage of genetic variation. The other percentage experiences genetic change without a 'start' with one individual. A whole group of individuals can experience a trending of change in a direction that makes it more compatible with the environment. Peppered Moths, geese and wild turkeys are not anomalies; they are simply three instances that have happened in the most recent 150 years. If it were possible to look at all creatures for 1,000 years, more cases would materialize.

Nature doesn't incorporate changes the way we breed dogs for long fur or size or color, by narrowing the genetic makeup. Several of today's breeds of dogs do, literally, have a shared mom dog in their past.[6] Shared-mom genetic changes go away when the species, no matter how unique-looking it has become, mates with others of its species again.

For the most part when instinct decides for a species that this change is a good idea, it implements it across the board and it is not lightly undone, certainly not with a little interbreeding. Even though the coal soot has been gone from the lion's share of England's trees for fifty years, there are still plenty of black moths around.

The Winter Moth

Possessing a will to live is true of all VS, of the mosquito, the mushroom, the tomato plant, the mole. This will to live is more appropriately called a will to reproduce. That the will to

---

[6] *I'm aware that I could use the legitimate term for 'mom dog' in this sentence, but I've lived with mom dogs and they shall remain mom dogs to me.*

live and the will to reproduce is one and the same is most obvious in short-lived VS like fruit flies, petunias, octopuses or spiders. They blow the wad on a one-time reproduction effort, devoting every bit of energy to it. Often, the individual could have lived another year or two had they not given their all to reproducing. Nature doesn't need them to live another year. Once the creature has laid and fertilized its eggs or ensured the seed scatters, their genetic code figuratively says 'you are dead to me.' Live, don't live, it hardly matters.

With our long lives, we have put a huge distance between the will to live and the will to reproduce. To us, they are two separate things.

Instinct can make some pretty odd choices when reproducing is the only important manifestation of the will to live.

In 2015 I became aware of the infestation of the winter moth in my town in Massachusetts when the moths dotted my home office windows after the sun set. In a few weeks the temperatures sunk to 15° F., and snow covered the ground, yet every evening the winter moths joined me, hanging on the windows that, due to double glazing, offered only light and no warmth. Why are they there? What do they eat? How can those paper thin wings and tiny bodies pump out enough warmth to keep from freezing, night after cold night?

I looked into it. The answer was: they do not eat. They have no intestinal tract. They never hunt for food. They can live up to seven months as a non-eating moth. All winter moths are males because the females have no wings. Both male and female do all their eating while they are caterpillars. After that, the male moth's sole task is to fly around seeking a wingless female hiding in tree bark to mate with so she can lay eggs. In the bitter cold.

Two questions arise: One, why does it have to be this way? What possible environmental conditions drove this species to be like this? Two, what was instinct thinking when it deleted the intestinal tract? Talk about burning a bridge!

The first question, how did the environment make this happen, is looking at it backwards. The environment doesn't 'drive' a change; the environment is the judge and jury after the fact. It rules on done deeds. It has no opinion on deeds not yet done. It is powerless to persuade or sway things in a certain direction. It only says aye or nay on things that are tried.

The environment has a set of rules, and it sticks to them. But it's not a long list. Creatures can do the oddest things that make no difference to survival. Wings can be all different shapes and sizes, needing faster or slower beating to keep airborne; ears can be more of a nuisance than help; predators can have long legs or short legs. A creature that's a worm and then changes into a flying moth for reproduction only in the winter was never required by the environment to do that. There was a space in the environment when it was less risky to fly around to find a mate, and that space was after the birds went south for the winter. No food to eat? The simple solution is to not count on finding food. Once a creature devotes energy to seeking out food and a big part of their body weight is the digestive tract, eating can become more compelling than reproduction. Looping back to the thought that in most VS the will to live and urge to reproduce are one and the same, eating is off-track. Ergo, deleting all the equipment for eating makes perfect sense when reproduction is the number one priority.

Whatever transitional stages, accident of birth or evolution took place culminating in a no-stomach, winter-thriving moth, if it reproduced, the judge and jury called it a win. As goofy as the strategy sounds, if reproduction happened, the species gets to do it again for another generation.

The environment doesn't 'make' any species do anything. All species simply try things; some work, most don't. Sexual pairing keeps most species stable and quiets variation. This enables instinct to reliably recognize food and predators.

Eventually a variation arises that helps the particular individual. Then basic statistical rules apply: if 80% of a species dies before reproducing (eaten, starved, no water, etc.) but only 70% of those with this variation die, and all survivors mate indiscriminately, within several generations those with the variation will make up 90% or more of the nearby population. It's not 'good' or 'best,' it just is.

In the case of the winter moth eliminating eating, we might think it short-sighted and doomed to fail eventually. We could say the same about the water lily choosing such a teeny-tiny niche. Genetic thinking might give anything and everything a shot, and the environment votes aye or nay once each generation.

Frogs

Nature prefers instinct to thinking because instinct puts an incredible amount of know-how into a tiny spot.

Nature can do wonders with instinct. Look at birds. Flying in a 3 dimensional world, adjusting to air currents, flying through the trees at 30 mph or more without crashing into anything, dealing with wind gusts, and landing on twigs that are bouncing around in the breeze. The quantity of split-second decisions are incredible. Mind-boggling, if it had to be done by thinking. To do what the tiny sparrow does every day we'd need years of training. In the meantime we'd crash into things, misjudge distances, and not correct fast enough to prevent crosswinds from throwing us into nearby objects. This inept flying would engage every ounce of our mental energy while doing it, at least for the first few weeks. Not so the sparrow or even the fly. They don't have weeks; they have to nail it in a day or two. And they do. No one teaches them; they instinctively cope with conditions materializing for the first time in their lives with all the skill of the oldest of their species.

Have you ever watched a fly and thought, 'Oh, there's a baby fly tooling around for the first time. Look how he botched the landing.' Nope. Because baby flies fly as good as old flies.

Nature may have taken 100 million years to accomplish it, but that doesn't diminish the point. When nature has her druthers, this is what she does. She stuffs 30 weeks of all-day training into a dot smaller than the head of a pin at birth.

All creatures on Earth that fly have instinct as their navigator. They are born knowing how to do it. Mom does not have to teach them. This is true of insects, birds, bats and even flying fish. Any ineptness or slow start is due to needing to develop to a size and musculature to fulfill their destiny to fly. It isn't lack of flying knowledge that causes slip-ups, it is strength and coordination, the bugaboos of all young creatures. Any thinking they do while flying is related to chasing food and watching for predators. While instinct controls the flying, their head can be focused on other things.

Frogs are a perfect example of the complexity that instinct can handle in a small package. Evading predators on ground and water, they adjust seamlessly between two entirely different environments above and below water. They know what to eat, what not to eat, where to sit, where not to sit, what's a predator and what's not. A frog does not learn these things; frogs as small as a fingernail simply know them once they grow legs. Given the complexity and difficulty of their lives, that only one out of a thousand lives three years is not amazing; it is amazing even one lives for three whole weeks. If they had to think, had to learn from mama frog, they wouldn't have enough time. They'd be dead.

Every spring in Massachusetts there's a day when at 10 PM the temperature is 38° F, or 3° C. Outside my office window I hear the frogs chirping their mating calls or territory calls or whatever they are doing. I mentally picture these tiny things, butt naked with their little soft skins, sopping wet, sitting in or at the edge of the little stream that flows behind my house, a stream that's merely a smidge above 33 degrees (1° C), a cold breeze wafting over them, and wonder how can they sing like they are happy? How can they be alive for even the next 15 minutes? They are not only alive, they're going to survive the night, and they are content.

Frog, resting on an instinctive non-threat, a turtle

The frog is the peak of evolution. It is the crowning achievement of life on Earth.

Not me, sitting in a 68 degree room feeling cold. Not me, who would be dead before midnight if I stepped out in my birthday suit to join them. Not me, who would never be able to figure out what to eat when I was three months old. Not me, who might not figure out how to mate unless someone told me.

Frogs in temperate climates actually freeze solid in winter, and thaw in the spring. Until a few decades ago people assumed that to survive the winter frogs dug down deep, below the frost line, even though their little soft paws did not seem capable of digging through two feet of hard-packed soil. Perhaps they borrowed old crawdad holes. Perhaps dozens of them gathered into a ball of warmth, with only the lucky few near the center making it to spring. Who knew what their trick was?

Youngsters would find frozen solid frogs and bring them home. Parents assumed that was the unlucky one, doomed. Into the garbage can it went. The next morning, surprise, it was hopping around! Everyone assumed it was found just before it died of cold.

Who could guess that frogs figured out a way genetically to freeze into a frog popsicle and then unfreeze itself later on?

Several years ago there was an article in a scientific magazine whose name slips me that detailed the chemical analysis in full monty scientific language to explain how frogs can freeze solid and not die. Wow, that seems like useful information, I thought. Space travel, here we come! Nobel prize for that lady. I resolved to read the entire article, looking up what I needed to look up to understand every word. I wanted to know if preserving people by freezing could be a thing.

The issue with surviving being frozen is not related to the speed or depth of freezing; the problem with freezing solid is that water expands slightly when it becomes ice. This is the sole reason ice floats. It spreads out a bit, so a square foot of water becomes 1.03 square feet of ice.[7] It's the same weight as a square foot of water but is 1.03 feet larger in all three planes. Each square inch has less mass than water, therefore it rises to the top.

Down at the cellular level that dinky expansion causes tiny cracks in the cell membranes because the cells are awash

---

[7] Ice expands 9% when it freezes. A 12" wide cube of water would become a block of ice 12.35" wide. Cube: 12 x 12 x 12 = 1,728 x 1.09% = 1883.52 square inches. 1883.52 makes a cube roughly 12.35 x 12.35 x 12.35 inches wide.

in water. At the moment when the expansion occurs, the organic cell membranes are colder and therefore less flexible than they are at room temperature. The scientific term is rupture. The cell membrane walls rupture. Not everywhere, not all the time; if even 4% of the cells have a single rupture, when the living thing is thawed, those particular cells don't work; their innards ooze out. It's usually far more than 4%.

This effect is noticeable in meat that has been frozen. There is a difference in taste and texture between fresh and once-frozen. Frozen meat can be a little bit mushier due to those ruptured cells. Nutritionally there is no difference.

Frostbite is an experience with freezing with which humanity is well acquainted. Whether toes and fingers can be saved is a factor of how long they were frozen, how quickly thawed (the quicker the better–but never use a heat level that would be annoying to normal skin. Lukewarm water is best), and whether the main body is capable of sending blood and repair materials to the site. Ruptured cells must be cleared out and new cells replace them. While this is happening, dead cells make the area look heavily bruised, even black. There can be lifelong residual effects from frostbite. Years later frostbitten fingers or toes may be quite functional but they chill swiftly even at room temperature.

Frogs, however, suffer no ill effects after months as an ice cube. In fact, they even revive themselves, at some point in the process restarting their own heart. Without losing even one toe.

It's a mind-blowing complex process. The changes occur at the molecular level and proceed in several waves. To make it brief and therefore inaccurate, it entails swapping out all the water in the body for an anti-freeze of the frog's own making (which I assume still has the necessary ability to carry oxygen and nutrients to all the cells) and then after chilling to a certain point, swapping that intermediate fluid out with 'something else' which freezes without expanding. When things warm up months later, the frog does not simply reverse the process. The frog initiates an entirely different chemical molecular process to come out of deep freeze and restore circulation.

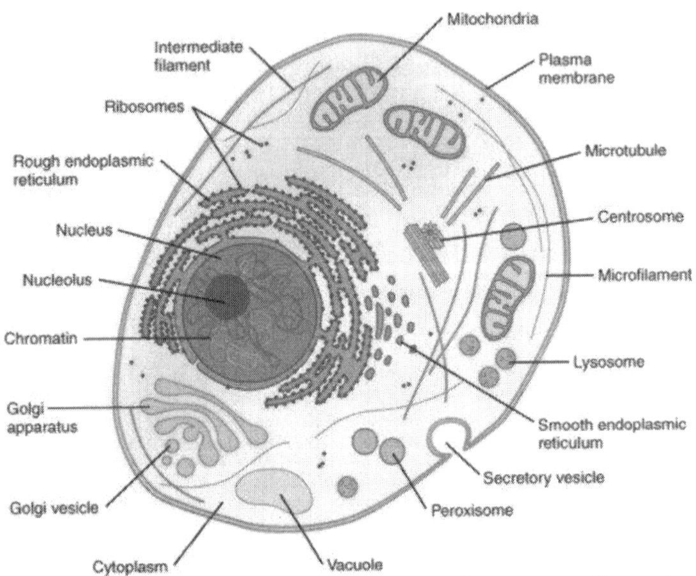

Typical cell in a multi-celled mammal, innards surrounded by a cell membrane. Most of our DNA is located in the Chromatin in the Nucleus, except for the tiny snip of Mitochondrial DNA (mtDNA) located in the Mitochondria. These little organelles convert food and oxygen from the bloodstream into energy for the cell.

Here's the creepy weird part; based on observations, it seems like a frog is able to choose whether to engage the thaw process or ignore a bit of warming such as a few 50° F. days in February. It could make that choice while it was an ice cube. It could sense time passage and grasp that it hasn't been long enough. As a human I will never wrap my mind around that. Current knowledge of how biological clocks work do not encompass entities that haven't had a heartbeat for three months. This is basically a rock with a sense of time passage. By every definition of the word 'dead' it is dead. But the frog is not dead in the frozen state because it possesses the ability to become undead upon engaging its thaw mechanisms.

For a frog, this process is an ordinary, every year thing. Show me a three-year-old frog in Wisconsin and I'll show you a frog that has done it at least twice. It's a transformation that defies belief. It is managed entirely by instinct and innate ability going right down to the cellular level. The only time brainpower has a part in it is when it seems impossible for the brain to be functioning. So perhaps that part is not judgment but instinct too. The idea that instinct functions when not a single part of the creature does is mind-boggling too. This could mean 'instinct' resides in the skin alone, the only part that would perceive warmth enough to initiate the reversal process before the tiny extremities begin to rot.

Most theories of life are rather broad and vague because those devising them want to encompass all the viruses, seeds and one-celled creatures with all their glorious oddities. But a frog isn't a virus or a seed or a one-celled creature, it's a walking around, singing, head-scratching being with eyeballs and fingers. It's doing something with which every single cell in its body–the eyeball cells, ear cells, intestine cells, muscle cells–plays along, even welcomes. In a way, every cell in the frog is nothing like our cells at all even though they perform the same function as ours. Long ago a frog's cells used to be like ours, and then they evolved into something far, far superior.

Instinct and its partner, innate ability, enables VS to respond to their environment swiftly and surely even when encountering the stimuli for the very first time. When that first winter arrives, the frog doesn't get panicky and think "WTF is happening? My environment is changing drastically!" Instinct has seen it before even if the individual frog, or even ten generations of frog, never has. Instinct knows what to do.

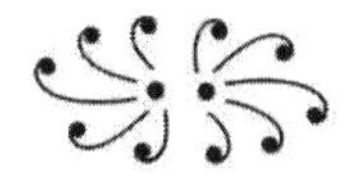

On a less fantastical note, have you ever seen a dead frog on the road? Did you ponder why this frog decided to leave the damp or watery place where it was living and head across a huge expanse of flat, hot rock, which is what a road is at the frog level? I have.

At first blush wanderlust is a very bad thing for a frog. There are more dead frogs on the road beside a pond, lake or water-filled ditch than in a dry area. One would expect a frog to hop out when its habitat dries up, but it's more common to see a frog mushed on the road beside an ample water supply. It had a good thing going. It left anyway. Doesn't make sense.

I grew up in a rural area of Wisconsin where half the land around was farmland and the rest was marsh. As civilization came to Oak Creek, roads cut right through the marshland in a grid pattern one mile apart. One day per year, in the spring, it was 'frog day.' Most people felt the same as we did; if we noticed it was frog day before getting in car, we didn't go out. That one day of the year it seemed possible that every frog in Oak Creek left where it was living and hopped around on the road. Fifty of them in every thirty linear feet of road.

It started in the late afternoon near the marshes, and the wandering frogs made it to half a mile from home by sunset. I can recall being out bike riding at the start of frog day, and pedaling like crazy to get home before it became impossible to avoid frogs. Several times I could tell when a horror story or sci fi writer has seen frog day; he or she uses the quantity of small, harrumphing, slowly mobile creatures as part of the plot.

As luck would have it, my mom was teaching me how to drive one spring day, only my second or third foray, and it became frog day. The shortest way home was through a two mile long stretch of road through a marsh. Most frogs are small. Driving over a frog bigger than a man's fist, however, that might be six or even fifteen years old is upsetting to a 16 year old. Through the car's open windows I could hear the big ones squish and pop. Or maybe it was bones snapping. I left carnage in my wake. Driving at 35 mph down a perfectly straight two-lane road with deep marsh on either side, I got the willies, literally. I freaked out. I wanted to stop but my mom exhorted me to keep driving. The only thing that kept us from going off the road was that my mom reached over from the passenger side and took the steering wheel while I thrashed about in my willies.

Wanderlust is a very bad thing for individual frogs. Why would they do it? Because the individual frog isn't calling the shots. It has no capacity to reflect, 'I've got it good here in the lily pads on the shore of his pond, I'm going to park myself right here, lay all my eggs here and never go anyplace else.' In twenty generations there would be so many frogs in that ten foot frontage of pond that none of them would get enough to eat. Yet the other side of the pond would have no frogs.

Since before roads were built, frogs celebrated frog day. They left their habitat and roamed. It was a feast day for birds and mammals. Frogs can outrun few, if any, land predators. Their sole protection on that front is to leap into the water. Yet at some pre-determined points in their lives they choose to take off in a direction away from the safety of the water and head cross-country, never knowing if they will reach water again. If they don't, they die trying.

Staying put is a wonderful strategy for individual frogs but is unsustainable for the species. The species must explore to see if another pond or stream has developed a half mile away. If nature allowed frogs to venture out here and there over the course of the summer, a far higher percentage would be eaten than if they all commence their wanderlust at once. Predators get a bellyful and stop eating. The frogs continue their wandering from late afternoon through the evening, when the hot sun is not another hazard.

Like a game of musical chairs, many of them find a piece of shore that other frogs abandoned. Perhaps some frogs that wandered off, having poor or perhaps no sense of direction, arrive at their same pond again. Frog wanderlust also broadens the gene pool, preventing siblings from interbreeding for several generations in a row. In this manner frogs can populate every suitable bit of real estate on the continent while stabilizing the species.

Why didn't nature put wanderlust into only first-year frogs? Why do all of them have to roam? I leave that to some other theorist.

Native Genetics

Genetic adaptation and genetic shift are common—but what is puzzling on the surface is the fact that different species have different quantities of chromosomes. Chromosomes are those worm-shaped things that come together in pairs, one from the male and one from the female in sexually-reproducing VS. All VS, even one-celled creatures, have chromosomes.[8]

Different species don't simply have different information on their chromosomes, they have an entirely different amount. Humans have 46 chromosomes, or 23 pairs. The horse has 64 chromosomes, the donkey has 62, the sunflower has 34, the carp has 104, both dog and wolf have 78, the opossum has 22, and chimpanzees, rabbits and potatoes have 48.

A species can reduce the count of chromosomes, as it appears humans did at some point after branching off from primates. Through all the pairings occasionally two chromosomes 'stick' to become one long one. The people who study this say it's easier for two to stick rather than one become two, because chromosomes are designed with end pieces called telomeres and a centromere that serves as a docking clamp during egg and sperm joining so the gene elements line up. Two inadvertently-joined chromosomes have telomeres on each end, plus two centromeres, so it joins up properly with the opposite sex's two individual chromosomes. Eventually, the one that stuck together becomes the norm.

---

[8] In fact, some one-celled creatures have several times more genetic material than humans. That doesn't mean our common ancestor had all that genetic material and our line lost it. It means that over the same period of time since then that one-celled creature tried and then turned off far more options than our line did on its path to becoming a large, multi-celled creature.

A species can increase the amount of chromosomes it normally has, but not by just splitting in two. It's a long description using many big words, and if you're really interested you can look it up on Wikipedia or Googling 'How can chromosome numbers change.'

Like any process done tens of thousands of times, mistakes are made. One case where we can actually tell when it happens is with Klinefelter Syndrome. Instead of one of her two X chromosomes going to each egg, one egg gets both X chromosomes; I assume the other egg is simply unviable. Both of them muscle in when joining with a male Y to produce an XXY offspring. If the man contributes an X, the offspring would be XXX. A Male with XXY might live a long and healthy life and in some cases have his own normal children. The double dose of Mom's X, however, has the effect of overwhelming whatever Dad contributed on his Y chromosome, washing out the effects of his natural testosterone. When this doubling up happens with one of the non-sex chromosomes it would be harder to detect. The same washing-out of the other sex's contribution would occur, but all we would perceive is that he is the spitting image of his father, or she has the exact same piano-playing talent as her mother.

At about the same frequency, 1 in 1,000, boys can be born with an XYY situation, two Y's from their Dad and an X from their Mom. If you start reading about this condition, you'll find the literature is a kettle of worms. The early genetic studies were performed on the easiest male population the lazy bum researchers could find: men in prison and mental institutions. Their conclusion was that men with XYY are found in prisons and asylums. Yawn. Everything else written after 1980 is bouncing off that theory, either to debunk it or confirm it.

To this day, if you tell a mother her son has XYY, she will cry. Some researchers say that XYY males are taller and more robust than average for the population, but others say the

XYY males are about the same size and body configuration as their XY male siblings.

Some studies show that school age XYY males require tutoring slightly more than average and slightly more than their siblings, or . . . it could just be that the mom, who obviously knows her child is being watched for mental deficiency, is fast to hop on getting a tutor as soon as the first D appears. I'm a mom and you betcha that is what I would do. The fact that a child has a tutor says more about the mom than the child. You go, girl.

This ability to sneak in an extra chromosome and overwhelm the other sex's contribution is very interesting. The statistic—and I'm not saying I believe it—that it happens with this one chromosome in two out of 1,000 births, half coming in on the sperm and half on the egg, means it could be happening with the other 22 pairs of chromosomes at a similar rate.

That could mean that 44 out of every 1,000 sperm-egg unions are stacking the deck to inherit features from one of the parties no matter what the other half contributes. From there it's an easy leap to imagine some of the chromosomes will activate genetic code errors floating here and there that were passed along as recessives for centuries, always being overruled by the other half so never manifesting until this happens.

Perhaps most of these end in miscarriages. When the doctor says the cause is unknowable and unavoidable, believe him. But sometimes . . . it creates blonde hair and blue eyes. By doubling down Or marries a cousin. Or both.

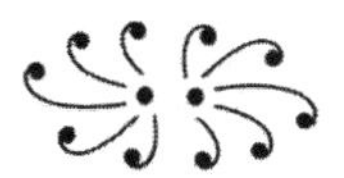

Instinct and the genetic thinking governing plants, insects, one-celled creatures and reptiles makes that VS very certain about what it needs to do to survive and reproduce. Any time 'thinking' is introduced into a species it is put there for the purpose of overruling instinct. Thinking is needed when an occasional modification to instinct is required.

Which is why moving-around VS will have an ability to think while anchored ones do not; instinct can make a lot of decisions, but it can't guide every physical movement in detail. Instinct can tell a VS to startle upon hearing a loud train horn 500 feet away, but after the creature hears it 200 times and nothing bad happens, only thinking can override instinct and determine there is nothing to fear. Brains can make a choice when instinct never encountered this fork in the road before.

Mendel's pea-plant and dog-variety model of genetics is useful for learning about genetics just like learning Newton's Laws is useful for understanding physics, even though his laws work only within a certain range of atmosphere, gravity, temperature and time. The variation-narrowing practice to change species is like Newton's Laws. It provides a visible and predictable result, but is not really how it works in the bigger world.

How does a species kick into high gear for genetic adaptation? What if 'kicking into high gear' meant a species suffered increased cell division anomalies during the sperm and egg creation if the VS was under stress or starving? This is exactly the case. From lab studies to real life, all living things have a greater propensity to genetic mutation when under environmental stress. For the most part these lead to miscarriages and deformities. Nature begins gambling when there is no sure thing.

Most pregnancies are stressful, but that kind of stress is normal. The word 'stress' in this context refers to environmental stress that impacts at the biological level. Significantly, the operative stress for female egg cell-splitting anomalies occurs 30 to 90 days prior to the act which fertilizes the cell. A poor diet during gestation can take a toll on the developing fetus, but it would not have a genetic impact.

One could argue that the mating of egg and sperm is indifferent to the emotional or nutritional state of the woman, since at that moment they are both independent entities. Like most things at the cellular level, it is far more complicated. There is evidence that the chemical composition of the fluids impacts the swimming capability of the sperm and the amenability of the egg, leading to certain characteristics of the sperm having a better chance than others.

There are also recent, very initial studies with mice showing that the flipping on and off of predator-warning genes may happen *in one generation based on only the mother's experience.* I shouldn't even be mentioning it because it needs to be repeated to be proven true, but if true, it would explain a lot about turkeys and geese.

In times of environmental flux and change, nature jacks up the incidence of genetic mistakes. Perhaps our rate of 44 double-downs per 1,000 is terribly high compared to other more stable species. One way to find out would be to pick some stable species—mice? Deer? Zebras? –and compare their miscarriage rates per 1,000 fertilizations to ours.[9] This would shed light on where we are, either transitional or stable.

---

[9] Sadly, what has too often been done in clinical and genetic studies is to test a tiny group of subjects and from that announce something happens once per [huge number]. I love statistics. But it's NOT statistics when someone looks at 53 subjects, finds one instance, and announces it happens twice per 100. Or once per 1,000. Both are made-up fantasy. It's appalling. To pin a frequency, the test group MUST be at least five times as large as the predicted incidence. Even that is weak. Those who come up with a frequency of one per double or triple their sampling size are LYING.

DNA and mtDNA

If you are well read you may have heard that mitochondrial DNA evidence is said to prove that all humans alive today shared a common grandmother, an 'Eve,' who lived about 200,000 years ago. The full story is easy to look up online. To assign the date, the proponents used the 'average rate of mutation' of mitochondrial DNA, called mtDNA for short.

When called to task on the use of the word 'average' in referring to the rate of mutation when the correct term should be 'our best guess at this time' since we don't even know IF there is an average rate of mtDNA mutation that can be relied upon, scientists back-pedaled on the shared-grandmother timeframe. They now give it a range of 400,000 to 95,000 years. Still, 200,000 years ago is the number that gets repeated all the time.

That makes it seem like DNA mutation is like a clock. It is nothing like a clock.

Before explaining why it's not a clock, you should know that mtDNA contains genetic instructions within the mitochondria that only the single-cell can use. It contains no instructions for the bigger creature. Scientists use mtDNA because it's easiest to collect from ancient cells, sometimes all they can get. No fault there. But creature DNA it is not.

It is DNA containing only instructions for the function of mitochondria itself. Mitochondria are the only organelles in a cell with their own DNA. Mitochondria function as the digestive system for the cell. Nutrients move from the blood through the cell walls and the mitochondria collect those and convert them into power for the whole cell. Representations of animal cells like the one on page 50 usually show a few mitochondria floating around, but a typical number is a few hundred per cell, or up to 25% of the cell mass.

When reproductive cells, sperm and egg, are made, each parent splits their DNA, sending along half of a DNA pair. Males manufacture sperm cells all their life. Females are born with a big stash of a few million eggs. Most of those fledgling eggs just die off over the years. One, maybe two, float down the fallopian tubes each month after puberty.

While the egg cell DNA combines with the DNA in the sperm, the mitochondria in the mom's egg go along for the ride as whole, fully grown little entities. Each time cells split during gestation, little mitochondria double themselves and each new cell gets an identical set.

A common-grandmother theory was published and even broadcast on PBS as true on the basis of mtDNA from 147 individuals. A similar study of the Y chromosome involving only 69 males was published as 'proving' the existence of a common grandfather, an 'Adam,' to all humans about 100,000 years ago. Later on another scientist found DNA from an African population which didn't share the common grandfather.

This is a perfect example of too-early results getting widespread ink which then has to be corrected over and over for a decade. Taking another 2000 samples from far-flung groups around the world is likely to find others without the Adam in their background.

While the recent common-grandfather story is debunked, there's about a 40% chance the grandmother one will keep standing. Treating it as a simple study in statistics, this sort of thing is likely to happen randomly. As with all statistical occurrences happening hundreds of times in a roll-the-dice fashion, once 'the house' wins more than 52% of the time, with enough occurrences it will win all of it.

Using mtDNA to arrive at a statistical majority is fine. But assuming it says something about the development of the larger creature is suspect. MtDNA could 'prove' a son isn't related to his biological father in the past 5,000 years.

That's why several years ago I made a frowny face the very first moment I read that important people in anthropology announced that on the basis of partial mtDNA, we shared no genes with Neanderthals (that were newer than when they split off 400K ago). I can see the Neanderthal in us walking down the street! I wondered how long it would take to back-pedal from that. Good men like Svante Pääbo, Director of Genetics at the Max Planck Institute for Evolutionary Anthropology, are doing their best to correct the genetic misinformation.

To keep the shared grandmother theory in perspective, even if true, it doesn't mean that the genetic pool narrowed down to one female, one 'Eve.' She had plenty of other relatives and friends around, and their children became the husbands of her female children. Five generations later a female would have inherited her mtDNA from her grandmother's mother's mother, along with a tiny sliver of her main DNA. All the other grandparents and great-grandparents would have provided main DNA too.

There's a deeper issue involved with leaning on mtDNA for information on the larger creature. Perhaps the creature DNA was changing drastically while the mtDNA remained unchanged, just chugged along. Conversely, nothing much could be happening between two population groups while by odd chance one of them had several mtDNA mutations in a short time. Therefore, even using the mtDNA as a timer for migration patterns is questionable; already there are situations where the 'timing' simply doesn't match the actual evidence.

We've had the tools to look at DNA mutation for less than 50 years, yet the current calculation is that one occurs every 125 years or five to six generations [early people lived on average about 35 years and began reproducing around 15]. This is a perfect example of the sample size less than the frequency.

One way this number could be validated is if we pulled mtDNA from perhaps 50 women who died around 1700 and then got genetic material from five to ten of each of their direct female ancestors today...but my guess is that the study would muddy the waters even more, making the clock thing impossible.

DNA studies are in a very early stage. Most of the papers have no real conclusions. They are just "I did this and this and this, and made a chart of the results. Period." You are not dumb if you can't make heads or tails about what was discovered; that tidbit isn't in there.

Although that can seem worthless, scientists know that each one is a piece of a 10,000 piece jigsaw puzzle, and we need at least 5,000 pieces to even start putting it together.

Even simply ruling out a causal link or a connectedness can be really important. Someday when this study is combined with six other studies, it will lead to a thought. But for now it's just raw data waiting to be used.[10]

We may never pin down a date of the existence of an 'Eve,' but I can't help imagining there was once a good mother who had a lot of daughters. Healthy, prolific daughters.

Human Tail

Every now and then, a person is born with a tail, or webbing between the fingers and toes, or hair all over their body. This happens because every single one of us has genes for a tail, for hair all over, and for webbed fingers. Over time

[10] Some of the most profound discoveries are dependent upon old studies where the researcher kept notes and data on minutia that was not deemed important at the time. We cannot know how even 'silly' research projects will prove vital. For instance, studies where we can revisit the same subjects 40 years later to get longitudinal results. These fledgling genetic studies may take on whole new meanings the day we discover what the genomes mentioned actually do.

we developed other regulatory controls that demand "suppress tail" and "suppress webbing." The command "suppress hair" may not even be a single one but a sliding scale or series of alleles that interact, and that interaction determines how much back hair or leg hair a given human will have. When scientists say we use less than 10% of our genetic code, it is because we have a heck of a lot of suppress and undo codes in our genetic makeup. The genes they foil are still there too. We have genes to do something and then other regulatory controls to negate that, so the sum is zero.

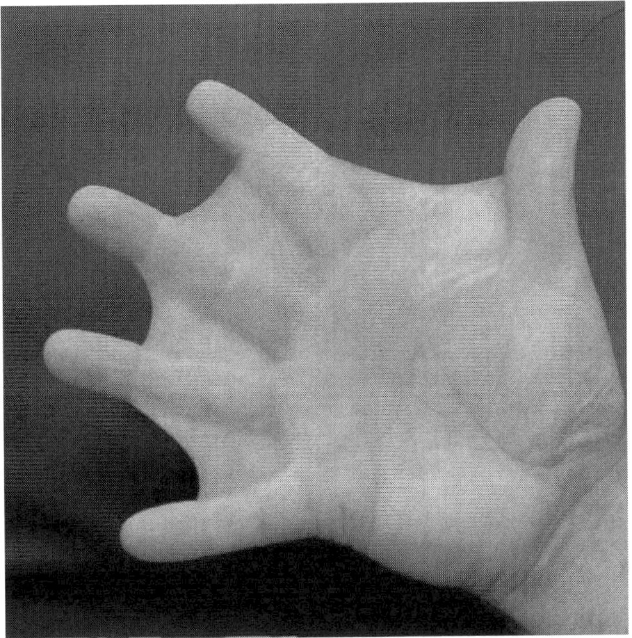

View of webbed fingers. Usually webbing is surgically removed just after birth.

When a baby is born with a tail, the defect was in the "suppress" command. A tail, with all the extra bones, ligaments, skin and hair follicles was not created from scratch, all that complexity instantly invented for the first time in a single generation. No, that's our normal tail. My 'inner tail' may be even longer and more wiggly than yours. When we don't have a tail it's because our suppress command worked. We did not have one of those random typos, i.e. mutations, in that spot. Human genetic abnormalities (and those of any species) do not run the full gamut of possibilities; many of them hark back to something that used to be a feature possessed in the past.[11]

In the case of a tail, it may manifest as a nicely formed extension of the tail bone. Webbing might be evenly formed and beautifully scalloped skin between all fingers. It isn't hard to imagine, however, that a gene suppressed for thousands of years will have mutated in that time. All genes experience random mutations. Those which cause something visible to change must either be useful, attractive, or at least tolerable or they end right there. When a suppressed gene mutates, nothing changes at all. With no outward sign, the defective gene spreads through a population. Only when the "suppress" control command experiences a mutation or several suppress commands combine in a certain way and let something suppressed get by, only then does the oddly-mutated suppressed feature appear.

Now a baby is born with something that was never any help to any past human. Most webbed fingers and toes manifest with errors.

Webbing, while it occurs about once every 3,000 births, is a dominant trait when it does appear. That means each sibling has a 50% chance of also having webbing, and when someone with webbing has children, about 25% will have webbing. Paternal genes carry the trait more often and it expresses with more influence than with genes from the

[11] Some birth defects are not genetic, but rather are a one-time occurrence: a partially absorbed twin, a bit too swift cell division at an early stage of development, leading to extra fingers or toes, or development stopped or hindered on some part of the body, often due to chemicals or other ingested materials, poor nutrition, or fetal circulation issues.

mother's side. Which tells us that our 'suppress webbing' commands are straining hard to hold the door against webbing on everybody.

Haplogroups

If you start digging into early man genetic studies, you'll find articles about genes referring to Haplogroups such as Halpogroup D. Researchers are focusing on specific bits of genetic code, noting mutations and tracking their spread by using both ancient and current day samples. They assume all mutations would be unique, so if two divergent populations have the same one, it must mean they're more closely related than others who do not have that particular mutation. But genes are not like snowflakes; they're more like a deck of cards.

In the majority of these studies you'll never find an answer to the question: what does that gene do? Often they don't know and seldom care. Researchers select a part of a gene in myopic fashion and track it through populations around the globe. They seldom, if ever, knows if it's active or inactive or what it does.[12]

Take the Haplogroup D research. Some articles are about the Y-chromosome [male] Haplogroup D, and some are exclusively the mtDNA Haplogroup D, which only comes from your mom's mom's mom's mom's mom, going back, you get the drift. No one pins a task on the Y-chromosome version or even knows if it's active. The mtDNA one just controls some ho-hum aspect of the mitochondria.

---

[12] If they are either using a sensible selection process or know what their chosen gene does, they don't say it in their reports. I'm on sturdy ground when I say if something huge goes unmentioned in an exhaustive, lengthy report, it's on purpose; they haven't a clue. That's indisputable for any scientific research.

One could not pick two more opposite things on which to pin the exact same name. One is only from Dad and goes only to sons, and one only from Mom but goes to both sons and daughters. If I ruled the world I'd say, 'pick some other Haplogroup to chase into antiquity, one that's shared by men and women and we know what it does.'

The researchers make complex charts of the mutations, so if an ancient bone has mtDNA version D2a1, then it has ancestors with D2a and D4e1. A find with D4b not only is much older than D2a1, it also split off long before.

I know that's all gobbledygook. It's made worse by not even knowing if this is an active gene or inactive. It would stand to reason that inactive genes get away with a lot more mutations than genes which actually do something important.

My gut tells me charting inactive genes is not the best use of our time. The output is starting to reflect that. Patterns observed in one Haplogroup charting are confounded by another. It's all predicated on an assumed steady rate of mutation and an assumption of total randomness in mutation, not that some kinds of mutation come easier to genes than others, which are the two weak bottom cards in the house of cards.

One ostensibly smart person concludes from Haplogroup charting that the Ainu of Japan, who arrived on the islands 15K years ago, are directly descended from Australian Aboriginal people. Look at a map to check out how many land masses they would have sailed past to land in the cold northern regions of Japan. Absolutely nothing else, from language to Ainu body features to physical evidence to Australian aboriginal sailing prowess points to an Australian connection. Sigh, Judge Judy rule again.

A far more likely explanation is the same mutation in the same spot on the chromosome happened in two places on the globe. Maybe that's about as random as two people meeting on a tour and having the same birth month and day. It's not common, but in fact it happens to you several times in your life. The only reason you don't know it is because you don't ask about birthdays with everyone you meet. When errors occur in the sequence of the four nitrogenous bases that make up DNA, Adenine (A), Cytosine (C), Guanine (G) and Thymine (T) it's likely but so far unprovable that some switcheroos are more prone to happen than others. Which would mean as you chase mutations of these sequences into history, you'll bump into, to use the colloquialism, twins from different mothers. Which is why one Haplogroup finds a match between the Ainu and Australians, while none of the other Haplogroups, or anything else, does.

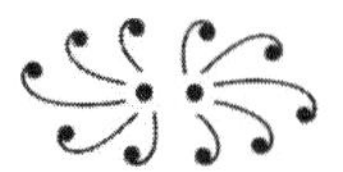

I am NOT one of those 'never trust nothin' on the internet' people. I love Wikipedia. I even donate to fund it. The solution is to read six things on a topic, not one. Often the first-page items in internet searches belong to those who paid money to SEO optimize; the real stuff starts at page 3. The better course of action is simply don't be a page-one addict on internet searches. Always remember the page one sites paid to be there; be cognizant that they have an agenda or financial reason for doing that.

Survival & Extinction

Nature, meaning life, reproduction and species does not measure success by how many VS are alive in a given year. Nor does it measure success by the amount of other entities that a particular VS competes out of existence. These snapshot measures are irrelevant to nature. The snapshot time could be one day, one year, or a thousand years; nature declares no one a winner based on that dinky amount of time.

There is only one measure that matters: can it sustain? Its behavior, color, evolutionary trends and habits must contribute to a balanced environment that can continue on for millions of years. If that happens, it wins. Being 'too successful' is failure. Wiping out its own food supply and eliminating long-time VS from the face of the Earth is an error, not an achievement.

Going the distance, survival-wise, is what life endeavors to do. VS that operate on mostly instinct, with few behaviors modifiable by using a brain, show a stunning laser focus on getting another generation launched. They don't eat, don't move, don't enjoy themselves unless it serves the need of aiding their survival until they are old enough to reproduce.

While VS living for one season give their all to reproducing, those living more than one year or having multiple breeding seasons must devote some energy to nests, hidey holes, and internal storage of nutrition.

Until humans started messing with them, the primary difference between annual and perennial flowers was that annuals bloomed all summer while perennials bloomed for two weeks. Annuals leave it all on the field, pardon the pun, because there is no next year to reproduce. This is it. Perennials, on the other hand, can pitch for their best odds but if nothing comes of it, they must retain strength and nutrients to try again next year. Which is why they often bloom by mid-summer and then spend the rest of the warm weather storing up strength to last the winter.

Today there are many perennials whose blooming season is several weeks long, and annuals that bloom even past the time when bees could reasonably be expected to be buzzing around. We could engineer that because inside the plant, just like it was inside the peppered moth and the turkey, there are vestigial switches. In the case of the flowering plant, they were there to shift the bloom season in either direction as well as change the color, smell and size of the flower. They exist because long ago the plant differentiated itself to secure a better niche by doing exactly that.

We give ourselves credit for engineering a flower that was only pink into white and red varieties. We can even engineer it to have twice the amount of petals, smell stronger and have yellow stamen. These can be done only because the flower has a history of modifying these things. Try to selectively breed a non-thorn plant to have thorns or a tulip to climb trestles like a clematis and you won't have any luck. Genetic selective breeding can't 'encourage' something that was never there to begin with. We can, however, get a plant to reach way, way back into antiquity to restore an old change or phase it went through. Often there are genetic remnants that are merely a 'more' or 'bigger' command regarding an aspect dearly involved in its survival. It might take 20 generations of selective breeding for the 'more' or 'bigger' combination of alleles to produce twice as many petals or twice as large a flower, which is a dinky amount of time genetically.

Selecting for traits that were never there, however, won't work in 100 generations or even 1,000 generations.

Species survive based on two different aspects that they own: one, they are as you see them, filling a niche in an adequate way with the looks, features and abilities possessed in the physical VS. Two, their genetic ability to adapt swiftly based on past precedent. Their genetics recall, for example, that within the past two million years ago the species was a different color, different height, tolerated cold better, didn't need so much water, had no thorns, and so on, and can pull those out of a hat at the least sign that the environment favors those again. Even when the 'environment' is only a breeder's hothouse.

Because the genetic memory exists in the whole species, the whole species can activate that switch in tandem when the environment changes. Past genetic stages or the 'more' triggers do not need to emanate from a single individual. Many flower colors were developed concurrently on different continents by flower aficionados who simply started with the handiest pool of twenty or thirty plants of that species and began selecting for minute differences. Their selection served as the environmental cue.

What's interesting is that invariably the new color, let's say white, was reached in very near the same amount of generations by those pursuing it with the same intensity. The one who started first was usually the first to get there unless his personal life interfered for a few years. Even more enlightening is the fact that brand new white-flowered plants bred by a breeder in the US could be mated with white-flowered plants independently developed in England and produce white flowers!

To make an analogy, it's as if the two plants reached deep for the same shared memory and once they had it, were identical again.

They did not change colors by different means, but by the identical means in parallel.

Neither flower knew there was another spot in the world where the same thing was happening; it wasn't necessary to know.

Neither flower quickened or slowed change based on what some other plant was doing.

Neither plant was cognizant that the perceived change in reproductive success was not happening to the vast majority of its species.

It only knew what its own environment was preferring. Each generation had more activation of the successful feature until a pure white was achieved.

This is how humans breed flowers, dogs, horses and cats to different colors, sizes and shapes. The only thing that prevents these members of a species from mating with other configurations is human intervention. These changes do not make different dog breeds into a different species no matter how different two members may look. This is because the changes are of the reach-back sort, not a moving-forward genetic change like that which makes cows and horses unable to mate.

The reach-back kind of genetic change, no matter how profoundly different from the current state it appears to be, can occur in parallel in hundreds of members of the species, or in members of the species separated by geography and time. The other, the moving-forward kind, usually culminates with a change in gene count of one of the branches. After this happens it becomes impossible for a sperm of one to fertilize an egg of the other, or the offspring are sterile.

Reach-Back & Extinction

A primary rule among scientists, even the past-studying variety, is that only the evidence that is collected can be used to create theories. No connecting the dots. So if VS that no longer exist are found in the geological record, they are declared extinct. It sounds plausible . . . but it's wrong.

We know species adapt to the environment, and can do so in such an eyeblink that it cannot be seen in the geological record; a species that reproduces annually can change significantly in 50 years. If we saw a 15" across moth in one strata and then in the next there were no more, only 2" across moths with a different wing shape, as things are today we say the 15" across moth became extinct and this new one popped up. Using the Judge Judy rule, isn't it more likely that the 15" one simply became smaller to suit a changing environment? We know that island mammals shrink in size in response to the smaller footprint; nature figuratively supposes it needs at least 100 of the species to make a good mix and this island can support only 80 of the former size but 130 of the smaller size, so smaller it is. Becoming smaller is easy. Because all large VS used to be small, a species only needs to reach back into its bag of tricks, not work forward. Conversely, very often 'larger' is not a leapfrog to huge but a command to increase a bit. If height or bright feathers or speed is valued in a mate or improves the odds of surviving to reproduce, the 'increase' allele gives each generation a sliver of advantage. In many generations it can double the size, amount of bright feathers, or swiftness of a species.

Flower and dog breeders know that most changes happen by selecting for incremental shifting in the desired direction. They are selecting for a stronger 'more' or 'hairier' or 'lighter' control command, not a change in the overall genetic makeup. Seldom do they have the luxury of finding pure recessive genes and popping out the change in one generation by combining two recessives. They have to work hard at assessing small shades of difference and breeding pairs of those.

Because up until recently breeding for traits was done using visually apparent changes, it explains why it is almost impossible for breeders to work on changes affecting only one of the sexes, that is, a sex chromosome trait. If they can't detect it in both the male and female, they can't breed to optimize it. Therefore, species engineered by human hands tend to have less dimorphism than the original species. When left to their own devices, male dogs will usually be bigger than female dogs. But many of our dog breeds today have almost no difference in size by sex; there is more difference in size between the largest and the runt of the litter, male or female, than between the sexes.

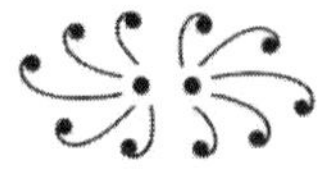

It is very difficult to tell from the stone strata what VS evolved into what a million years later. There is a place that retains the story: the genes. The next chapters of paleontology will be written by geneticists. They won't need to find old DNA; the story is contained in the DNA of VS living today. It's just a matter of time before we can decipher it.

Recently we've collected enough intermediary stages on several large species to string them together in an evolutionary lineage to beings alive today: the horse, the whale, us. Those earlier versions of horses didn't become extinct. If *extinct* is defined as running out of breeding pairs so no more are born, then that didn't happen to those horse ancestors at any step of the way. They look different than horses today in the same way you look different than your grandparents. You are probably several inches taller than your great-great-great grandparents, but that doesn't mean all those short people from 1810 are extinct.

The evolution of horses from a dog-like creature with paws to a hoofed large animal.

When we look at the small animal that was the ancestor to today's horse, or the land creature that became a whale, it's obvious there were transitional stages that were not as well adapted as today's creatures are. Imagine all the phases between having claw-like toenails to having hooves, which are enlarged toenails suitable for running upon. As jaws changed, there were probably stages where the creature had significant tooth problems. As tails changed, there were stages when the tail was barely adequate for the purpose the creature needed.

The problems may have inconvenienced the living VS but did not impede reproduction. Like humans, many of the worst effects of being a transitional creature appear towards the end of the breeding age: bad backs, sore feet, poor hearing, skin conditions, knee problems. Instinct doesn't care; once the VS reproduces and gets the offspring off to a good start, it's not genetically selecting to eliminate those things. When was the last time you heard of a woman refusing to marry a guy because both his parents had knee problems?

The Problem with Finding Bones

Ancient bones can be found only if there was an exceptional circumstance surrounding their death. Certain physical environments will never, ever give us an old bone.

A rain forest will not give us an old bone. When a creature dies in that lush woods, after the birds and small rodents had their fill, fungi and bacteria will take the rest down to a microscopic level.

Normal predator kills will not either: several levels of other creatures, culminating in the roots of plants, will nourish themselves off the remains. It stands to reason that when times are good, every single instance of a species will be fully returned to the earth, in one manner or another, within 100 years. After death, 99.999% of all living things will be indiscernible in 300 years.

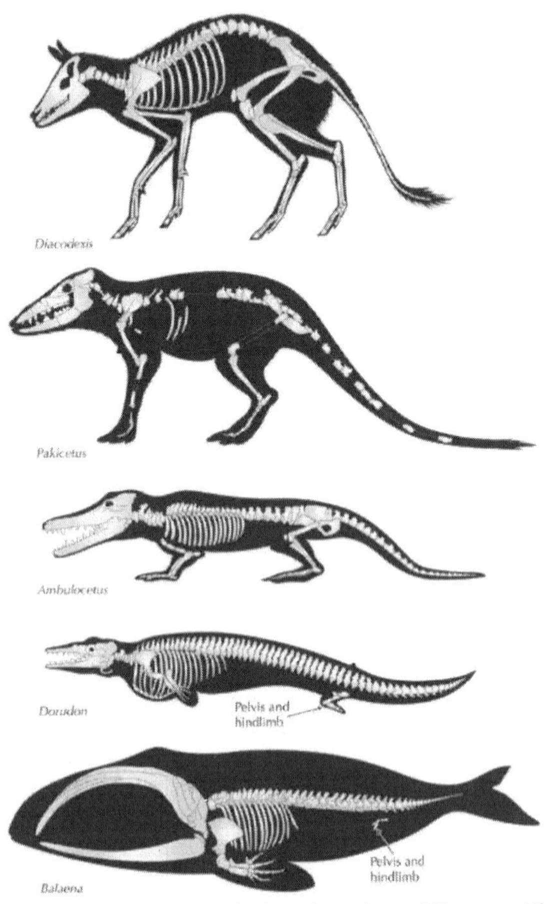

A series of fossils showing how whales adapted to a life at sea. The whale had its tail turn into rear fins, but the seal transformed its hind legs instead.

We already know the evolution story for several mammals using both the ancient bone record and modern genetics. Current theory is that zebras and donkeys diverged only one million years ago, while their common ancestor diverged from the horse two million years ago. Today the zebra has 44 chromosomes, the donkey 62 and the horse 64. This highlights that it's not distance in time or changes in physical features, but chromosomal count changes that nail down separate species.

Today's Phylogenetic tree (families of animals by their genetic relationship) shows that horses and bats had a more recent common ancestor than horses and pigs; that humans and rabbits had a more recent common ancestor than humans and sloths; that elephants are equally as distantly related to hippos and cats as they are to humans. For a fun look at who is more closely related to who, visit http://genomewiki.ucsc.edu/index.php/Phylogenetic_Tree

When it comes to humans and our nearby simian relatives, the differences in estimates, which range from four million to ten million years ago, hinge on estimates of genetic mutation rates. Assuming a steady and longer distance between mutations puts our monkey past further in the rear view mirror. If mutation rates are responsive to environmental stress, such as that created by glaciers during an ice age, it stands to reason that mutation rates increased in a huge fashion on both ends of each icy event; when dropping temperatures changed the plant life and allowed glaciers to cover 30% or more of the surface of the Earth, and then again when things heated up, once more playing havoc with plant life.

Counting just the outs because the ins would be gradual but the outs catastrophic, these events happened most recently 420,000 years ago, 190,000, 140,000, 80,000 and 12,000 years ago.[13] This is not counting four others that happened between 1.5 million and 560K years ago.

Calculating faster rates of mutation for the 3,000 years around each environment-changing event could leave us

[13] These dates are always changing and vary crazily depending upon the source. There is no agreement on the wax and wane of any glacial episode within 10,000 years. While humans have not nailed down exact millennia, we are certain there were many wax and wanes and that the last one cleared out of the northern hemisphere between 12,000 and 14,000 years ago.

looking at a divergence more like five to four million years ago. This more closely aligns with the fossil record.

Features taking a million years to evolve in species will have intermediary steps that are never found because none were memorialized. The worms ate them. Theories based on only what we've found to date is bound to generate errors. It makes a laughing stock out of science when the next find comes in. In that light, speculating using the Judge Judy rule is on firmer footing than basing theories on each individual finding, ignoring what we know about life and biology and the kinds of adaptations that go with specific environmental conditions.

Developing a good understanding of instinct's well-worn grooves and using those grooves to develop theories, or more importantly to hold off on premature theories, is likely to get us closer to the truth. Even as I research for this book I find articles dated 2013 that state things already no longer believed in 1990 by the pertinent scientific community.

Mistaken theories about how the native populations got to the Americas, the differences between Neanderthals and modern man, how people got to Australia and Easter island, and several others linger about for decades after being disproven.

There is some evidence that dolphins are second in line in intelligence, behind us. Being a species with flippers so owning no propensity to manipulate items means they won't be carrying around spare food in bags or making necklaces. Their speech and communication may prove to be instinctive, not learned. What is intelligence, exactly? It's an ability to replace instinct with learned adaptation. You can feed a pet snake for five years and it will never learn you are a friend because its instinct says you are not. If it is hungry or stressed it will do what snakes do. On the other hand, you can feed hummingbirds from a feeder on your deck and in a year several of them will recognize you as a friend and with practice and patience even feed from a vial you hold in your hand. Learning can overtake instinct in an individual member of the species.

I used these two creatures to make an important point: brain size alone doesn't reveal the capacity of a creature to learn. A hummingbird's brain is far smaller than the brain of a large snake yet it can modify instinct while the snake cannot. Many scientists propose that it is the proportion of [the cognitive part of the] brain size to body size that indicates intelligence. This is half true; there are notable exceptions. That theory begs the question: if brain size = intelligence in tandem to weight, why can smart small creatures pack twice the brainpower into each square inch? That must mean brainpower has a minimum package size much smaller than we like to believe.[14]

VS with intelligence also have instinct; individuals have the ability to assess their experience and modify or even ignore instinct. The ability to think evolved to improve adaptation to highly variable circumstances. As a creature divests more and more of its behavior to thinking, at some point instinct becomes so weak or inadequate in a few regards that the young are at peril to learn on their own.

To compensate for a good-idea strategy that has now come too far and is starting to be a hazard in the first weeks of life, another change is stacked upon the first. Where before a parent may have merely stood guard over eggs or the very

[14] The size of our heads is most likely the result of cultural selection for largeness over the generations. Reproductive selection is the cause when a physical attribute grows large without a survival-related benefit. For thousands of years people have been puffing up the size of their heads with hairstyles and hats. Some cultures bind the skulls of babies to make the head taller or longer. Men dislike baldness because it makes their head appear smaller, ruining reproductive success. Awareness of this led to religious head shaving as a sign of humility and head shaving as a punishment.

young before, now the young must also learn some things from the larger animal.

The most basic version of this is a behavior that evinces in only a small window of time: imprinting. Baby ducks and geese are a few of the species that 'imprint' themselves onto the first large moving-around creature they see and imitate it. In the wild it is invariably a parent, but with domesticated birds it sometimes is the family dog or a person. A young duck in this situation will show no desire to swim and will make champion attempts to eat what its new parent eats, sleep with them, and follow them around all day. To a lesser extent dogs and cats are similar. If people handle them in a friendly manner before they are four weeks old they shall become what we consider normal. If there is little handling or playing until after 8 weeks, it will take a much longer time for them to warm up to all people. They may permanently remain a 'one person' dog or cat.

Does a duck that imprints on a person or a dog remain like that their entire life? We know that answer; it does not. That behavior gets it over the hump of being young and inexperienced. Later, even with an early experience to the contrary, it finds out it likes the water, it likes to fly, and it will figure out what it likes to eat. Being able to learn also means having the mental power to reject learning, or change an opinion about what was learned.

We are used to the notion that baby animals 'love' everybody they meet. We know that even the meanest mammal alive is cuddly and friendly at the baby stage. Some humans think if we love them enough they will grow up to defeat their instinct. And they do, for the most part. For 99% of the time. It's that 1% when they forget, or are stressed and listen to instinct, that disappoints us.

Intelligence up to a point is a successful survival strategy. Predatory creatures tend to have more intelligence than prey species, but that isn't a hard and fast rule. Even though quite smart predators have developed such as wolves, dogs, tigers, cougars and ferrets, there are still plenty of mostly instinctual ones around: alligators, lizards, crocodiles, sharks, frogs and snakes, to name a few. The whole insect world is instinctual.

It's not clear that there's a definitive long-term survival advantage for intelligence over instinct. When something unusual happens, creatures with intelligence find it more possible to get out of the way of impending doom than instinctive ones. Instinctive creatures count on merely being spread so far and wide that any localized disaster will not endanger the species. Which is the case most of the time. It is in the moments of a creature's life when most of the time doesn't apply that intelligence has a leg up on instinctive.

The Language Inventors

We are not the only VS on Earth with language. Bees have a language. Individual bees fly off in exploration. When one has found a rich source of pollen, it returns to the hive to perform a dance that gives the directions to other bees. With the directions they can fly straight to those flowers even if the original bee who provided the directions doesn't come along.

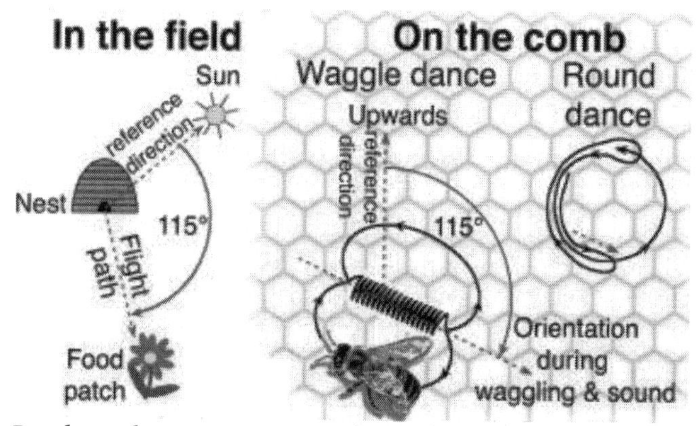

Bee dance language, portion that indicates direction. Other dance moves indicate distance.

Bee language is not taught to baby bees. They are born knowing their language. Because of that, the amount of brain it takes to master and hold all the nuances of their language is miniscule compared to us. Don't sniff at their language; it is comprehensive enough to give direction, distance, type and quantity of source. For all we know there are moves that provide information on size, 'this is a 10-bee job' and obstacles 'fly high along the cliff, there's a big spider web near the ground.' Or maybe not.

Imagining aside, this is how instinct usually handles language, by implanting it. All dogs have the same language even if they haven't seen another dog since puppyhood. Watch Cesar Milan interpret dog interactions, spotting communications that elude most of us, and you'll catch that it's an absolutely unambiguous language to them. Who sniffs first, what ear position means, and how the intent of a gesture changes depending on what the tail is doing or where the dog is looking communicate very specific messages every dog understands. Cesar can predict the next 'statement' that will be made. Dogs can say 100 things to other dogs. Humans can learn the language, but dogs are born knowing it.

Until we can prove dolphins have regional dialects and idioms, dolphin language is probably instinctive too.

Not so with people. We have an 'invent language' gene. We don't just learn language or use language, we make it up. Every little puddle of people starts modifying the language until it's barely understandable by the people of nine generations ago.

The amount of brain space that it takes to master language is significant. There's some evidence the location spreads out a bit, and that languages with a lot of vowels like Italian, Hawaiian or Japanese are stored in different locations in the brain than languages with a lot of consonants like German and Mongolian. What this means is unknown, but it is interesting.

We will make up a language from babyhood if no one tries to teach us. Or even if they do try. Twins sometimes create their own language; it's even more common among triplets. Years ago I knew two guys who were in their early 30s and working as machinists. They were part of a non-identical set of triplets, two boys and a girl. One was a friend of mine and in chatting he mentioned they had a hard time of it while young. The three of them so preferred their own language that their mother couldn't get them to speak English sufficiently enough to go to kindergarten at 5. The only way to get them fluent in English was to place them in different schools, two private and one public.

I could go on and on about what that quintessential story says about learning from peers vs. learning from parents, and that lapses in parenting isn't the cause. Their parents obviously cared a great deal since the solution was pretty expensive and time-consuming. Later I found out their sister also worked in the same plant, in the lab. Sometimes I would see the three of them eating lunch together by the windows. I itched to walk by to hear if they were speaking their own language, but let them be.

We love learning language but along the way we invent pet names for loved ones, use words in new ways and create words that have meaning only to our family or our best friends. Our lives become a source of paint for the canvas of language. We take the funny mispronunciations of babies and adopt them into our everyday language because we find them amusing and endearing.

Because of our 'invent language' gene, language evolution behaves the opposite of genetic evolution. When a small population is isolated from the main population for whatever reason, in genetics the small population will move away from the larger population's norm; the large population may not change much in 300 generations, but the small population will almost always have some differences by then.

With language, the bigger cluster changes faster than the smaller cluster. Researchers have said that the English spoken in Appalachia in the 1920s was more like the rural British of the 1700s than folks in rural Britain spoke in 1920.

To exaggerate to make the point, if every person alive adds or changes ten words in their lifetime, then the language of a large group will change more profoundly in a few generations than the language of a small one. We can cope with this because language is our sea and words are mere ripples. We learn new words our entire lives.

Using the latest idioms is something we do delightedly, not because someone forces us. A few years ago my son spent a minute or two teaching me how to properly use the idiom

"....said no one ever." He gave an example or two of improper use, including a TV commercial, which he snorted was obviously written by an old person who didn't know. The glaring question is, why learn the rules around a phrase when there are other equivalent ways to say the same thing? All I can tell you is that I was raring to go and on the hunt for any situation where I could employ my new phrase.

Six dogs can hang together for ten years and never invent a single word among themselves, even though they can be taught the meaning of a few dozen human words. Six humans can't hang together for even two weeks without coining some word, phrase, or inside joke. Most of the time we don't make up words; we borrow from other languages, borrow from sounds we hear, stick two words together or repurpose a word that used to mean something else. We are so loosey-goosey with language that it's a wonder people in California can make out a single sentence I write. If you traveled 400 miles in any direction a thousand years ago, good luck with being able to converse at all.

Selective Sweep

Did both early humans and Neanderthals speak? Yes. Both had a hyoid bone that was shaped to assist the complicated tongue, pharyngeal and laryngeal [throat stuff] motions involved with speaking. The U-shaped hyoid bone serves as an anchor point for these motions. Other animals have a hyoid bone, but they are smaller and located below the larynx. At birth our hyoid is below the larynx–this allows babies to nurse and breathe at the same time–but at about three months the larynx starts to drop. When the larynx drops below the hyoid, the fine muscle control that's necessary for speech is possible. Both size and location of the hyoid bone determines speech. Only humans and Neanderthals have it in the right spot for speech; in every other primate and animal it is too high in relation to the vocal cords.

Scientists have identified a gene heavily involved in speech, oral and facial muscle control. If this gene suffers mutation, it impairs the physical creation of language. It's called the FOXP2 gene, and it is found in both human and Neanderthal genes but not in other early man branches.[15] Once the FOXP2 gene showed up there was an extreme natural selection for members having the gene. It's called a 'selective sweep' in genetics.

Selective sweep means that reproductive success was so tied to that feature that pretty soon the entire alive population had it. It means someone with the feature would have an extreme mate-getting advantage. This is a good place to note that it's not always the environment that causes genetic changes. Monkeys with red butts and peacocks with their fan of tail feathers are examples of attractiveness to the opposite sex causing physical changes.

Whenever a selective sweep occurs, neighboring genes tend to homogenize. It creates a bit larger gene section in which both alleles are identical. Pulling along neighboring genes is scientifically called 'hitchhiking.' Why make up a new word when an old one will do.

Humor aside, it appears language was very useful the instant it showed up. The trait is universal in humans and in Neanderthals too, as far as we know right now from a handful of Neanderthal DNA samples from caves dating between 45,000 and 36,000 years ago. As of this writing we don't have

---

[15] Whatever date becomes nailed down for the divergence of us and Neanderthals, even if it's not 400K years ago but the previous ice age 600K or earlier, language occurred before that. Genetic diversity of Neanderthals was 1/3rd that of the concurrent human population, which supports the theory that all the Neanderthal bones analyzed up to now are descended from a single small family group that headed north.

a complete enough DNA profile of the Denisovans to confirm that the FOXP2 is there also.

Denisovans

Who are the Denisovans and why have you never heard of them? Mostly because we 'found' them only around 2010 via DNA testing. Current thinking–and it will change–is that after our common ancestors trekked to Europe 400,000 years ago, a few thousand or tens of thousands of years later the ones called Neanderthals stayed in Europe and the ones that headed to the far east became Denisovans. The two branches were cut off from each other. Each developed a little differently but they shared more features than they did with humans. Denisovan genes have been identified as the ones that permit humans to live at high altitudes with less oxygen, be more resistant to cold, and heal faster after cuts and wounds. As more genetic material is analyzed and compared to humans today, more will come to light.

These two branches were the only inhabitants of the northern hemisphere[16] until after a few more waxes and wanes of glaciers, when they were joined by homo sapiens heading out from Africa 60K years ago. In the intervening time the humans had changed too, because their climate and diet was different.

All of today's humans are descended from that African branch of 60K years ago. There was interbreeding as homo sapiens traveled and settled over 30,000 years. The bones of

---

[16] Homo Erectus is believed to have headed out to Asia 1.8 million years ago. There's no evidence yet whether this line looped back to contribute to human DNA, or simply became different like donkeys and zebras did. Co-existing is not meaningful in itself; many species of monkey co-exist, as do many grass-eating animals, in the same territory. Current theory is that Homo Erectus lasted to about 40K years ago, so could have met the Denisovans and even humans. They didn't speak.

one long-ago homo sapiens just recently DNA-analyzed revealed that a great-great-grandparent had been a Neanderthal!

In Europe, individuals to this day may have up to 2.5% Neanderthal DNA, while Eastern populations and Pacific Islanders may have up to 5% Denisovan DNA. While that seems tiny, they could be very active bits. Humans, as well as apes and mammals, use around 8.2% of all their DNA. No one assumes all of the Neanderthal or Denisovan genetic material is active, but the percentages reveal just how much of our appearance, diet and features could be affected by a relatively tiny bit of our genetic makeup.

FOXP2

A date has not been pinned on when the FOXP2 gene showed up and which years the selective sweep was taking place. There is, however, consensus that both Neanderthals and humans spoke language.[17]

Speaking is not intelligence, and intelligence is not speaking. Many creatures have a language of sounds and gestures. Some are quite sophisticated, like elephants and whales. Speaking, however, would make instruction of the young easier and would beef up memory. Memory is not intelligence either. Yet what we consider a poor memory is spectacular compared to other primates. We can recall the

[17] Prejudice is still a strong force in Neanderthal characterizations. Initially, DNA researchers announced unambiguously that we shared not a shred of genetic material newer than our 400K branch-off date. Their human comparison group was 53 individuals. Neanderthal genome strings are now used to identify shared genes less than 60,000 years old. Several traits common in non-African populations, like freckles, light skin, eyebrow ridges and straight hair are strongly indicated (not proven!) to stem from the 0.05% to 2.5% of Neanderthal genes in Europeans and the 0.05 to 5% Denisovan in Asians.

words and grammar of whole languages, sometimes two or three of them, plus a few hundred TV ad jingles and the names of our school teachers thirty years later. We remember street names and directions to eighty different places. Even 1/10th of this is beyond what any primate could muster with training and help, and most of it we retain without even trying. Intelligence without memory means reinventing the same food-gathering, shelter-building and storage tricks every few generations. This is exactly what our history looks like for 3 million years, with advances appearing and then 1,000 years later they're gone, not reappearing for 30K years or more. We know memory appeared late in human development, lagging a proverbial three steps behind language development, by simple deduction; remembering would cause a learning curve over generations. We don't see that; we see advances over hundreds of generations, at a pace even instinct can manage.

Inexplicably, the genetic experts identifying the astounding selective sweep around the FOXP2 gene act coy on whether the speaking gene was the cause of the selective sweep or simply a hitchhiker.

You tell me. (see what I did there?)

The Water Monkey

"No man's error becomes his own Law; nor obliges him to persist in it."

-- Thomas Hobbes, English philosopher (1588-1679)

It's repeated over and over so it must be true; early humanoids left the trees and began walking upright on land so they could hunt big game on the savannah, across the open plains, in the fields and grasslands. Museum dioramas show hairy early men with clumsy neck jewelry and loincloths shaking their pointed sticks menacingly at a huge lion or other sort of oversized interloper, scrabbly grass and big sky in the background. Nearby, half under a rocky overhang, females crouch over a bit of an antelope they are preparing for a meal. The caption says early man hunted big game and walked upright to see over the tall grass and stalk prey. We lost our hair because we needed to keep cool during long runs pursuing game in the open plains.

Never mind that our bodies, teeth, hair, ears, nose, stomach and metabolism make no sense for that environment, based on what we know about the plains environment and the large mammals which live there. If we'd been adapting to hunting in the high grass open fields for the past five million years, we'd have acquired the traits useful in every single land predator and savannah dweller.

What a living thing does not do after five million years is as significant a choice as what it does do. What we don't do well is tolerate being in full sun for hours every day.

Not a single open field predator or creature has thin-skinned bare shoulders. Not now, not in the past, ever. Reptiles, toads, worms and other bald creatures that may live in the open field environment are shade-hugging, lie-in-wait creatures, often nocturnal or at best feeding around dawn and dusk. Armadillos and rhinoceros developed thick, hard hides. We're not simply a creature that gets sunburn; overexposure to sunlight, regardless of skin color, leads to a high incidence of pernicious skin cancer.

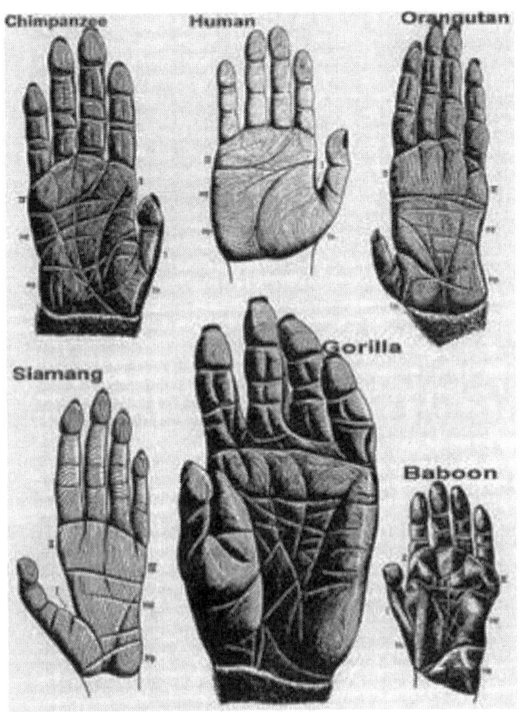

Generalized hands of several species. The hands shown here are all laid flat for comparison purposes, but while living some of these species would find this a very difficult pose to hold. They would manage it only by pressing their hand to a flat surface, not by holding their hand vertical in the air.

Our own bodies tell us we are newcomers to the big sky life. We get dizzy and faint when exerting ourselves in the hot sun too much; we need to wear a hat. There's a reason why baseball players wear a baseball cap and it's not team spirit. Wherever we look on Earth, and however far back we look, farmers always wore big hats while working the fields. If we had even a two-million year heritage of trotting around in the noontime sun I think by now we could skip the hat.

Our hands are different from the always slightly curled nature of the tree-climbing common ancestor, but in a strange way; they became flatter. The palm became wider and the fingers became shorter. The nails became weaker. If our hands were used for gripping tools and holding onto things pretty much like those of our tree-climbing ancestors, retaining the handy curl-around feature would have been a plus. If we were predators that stalked and defeated our large-game prey, tougher nails would be a little helpful.

Another aspect we don't share with apes is that our fingers and toes shrivel when wet. Mark Changizi, an evolutionary neurobiologist and author who is more famously known for inventing glasses that correct colorblindness, performed tests and found that the shriveling acts much like tire treads on rainy streets; it improves grip on wet objects. Medicine has known since the 1930s that it is an autonomic nervous system response to water immersion. Areas of the hands and feet that suffer nerve damage will not shrivel. It is a human adaptation to handling items while submersed in water.

Looking at our feet, it's apparent they are in a barely-OK transitional stage from a branch-holding foot to something to be walked upon. In transitions, nature does what it can with the raw materials available. For instance, in the lineage that became whales, the tail was converted to a tailfin, while in seals the hind legs were repurposed to do the exact same thing.

Look at a drawing of the foot of a well-adapted tree-climbing ape and our foot beside it. One of the most profound differences is that our 'heel' portion narrowed up. Horses, dogs, cats, camels, and most (perhaps all?) savannah animals consistently show a shrinkage of the feet until they are directly below the leg. To oversimplify, their feet have become just a knob below the ankle. Our feet seem to be going in the opposite direction. Our heel has gotten skinnier and we developed a wider, splayed part near the end, several inches from the ankle. The toes have shrunk to uselessness.

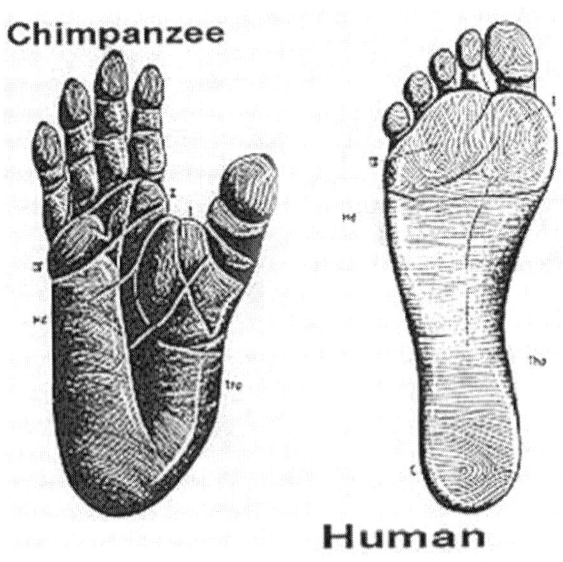

Comparison of feet. Note direction of evolution.

After three million years of chasing big game, humans became the fastest two-legged creature on the planet–said no one ever.

There is hardly a walking-around creature alive that is worse at running than ourselves. Almost everything over 15 pounds can outrun us, and a great deal that's smaller too.

With three million years to work on improving our pursuit of game, one would think that was enough time to do better. Obviously, we were developing this ability to stand while getting better at something else entirely. Our legs and hips look like they may have been adapting for upright walking, but our feet say no, absolutely not.

Logically, early man, with their feet in even worse transitional shape than they are now, didn't chase down and catch big game with their swift speediness, their mighty strength and their nasty fatal bite.[18] Because feet today are even more perfected for walking than they were 800,000 years ago, we can run on two legs faster now than ever.

Foot evolution came with another change that our simian relatives do not have: we can point our toe. When we retain our babyhood flexibility, all human beings can more than point the toe, we can overarch it, meaning make the toe go past the centerline of the leg bone. This trait isn't useful for running, which uses the ball-heel action as a spring. Our foot can flop back and forth just like a flipper when we swim.

When a foot is viewed as a swimming aid, the shape of it suddenly makes more sense. Getting wider farther from the ankle acts as a paddle. A smaller heel and more flexible ankle improves propulsion. When humans want to swim faster we strap on flippers, which have an uncanny resemblance to the way our foot would look if we used computer software to measure the changes from the foot of our early common ancestor to today and extrapolated the trend out a million years. No software starting with our common ancestor's foot and morphing to today's foot would shoot out into the future and come up with a good plains hunting-animal's foot.

In past chapters the case was made that when performing the same task, the affected body part of the bird, mammal, marsupial or reptile will begin to resemble each other, taking on the optimal shape to perform that task. With five million years to do this in the plains, and with the added boost that women prefer to mate with good providers, our foot should

---

[18] In fact, grown men who work out are no match in hand-to-hand combat with the average 80-pound female chimp who doesn't work out at all. It's indisputable that humans have gotten physically weaker since branching out from our common ancestor. Even caged chimps weighing under 100 pounds can pull several hundred pounds on a dynamometer, and a chimp on all fours can outrun a college sprinter. We've become *less fast on land* and *less strong* than we were before we began attempting upright walking.

be well on its way to looking like a paw, having large hard nails that dig in to facilitate a sprint. Instead, we have this soft, floppy thing that merely develops callouses if used too much. I admit, however, that no compelling case for our being an ape on the path of turning into a dolphin can be made using only hands and feet; there must be more.

Take the circumstance of our ears. Humans have an ear wax problem. We pump out too much. Since ear wax collects dust and bugs, well-adapted land animals have a slightly oiling coating at most in their ears. Dry as a bone is normal for field animals.

Why do we have ear wax? There are functional differences between the ears of tree dwellers, savannah dwellers, and water-immersing creatures. Tree dwellers tend to have ears like cups, large and on the sides of their head. In their world sound comes not only from sideways but from up or down, so they swivel their heads to locate where a sound is coming from. It does no good to just hear a danger and not identify it.

Savannah and plains animals like wildebeests, bison, rabbits, deer and horses tend to have longish or oval ears near the top of their heads. Their ears swivel independently to pick up sound. Danger is in a single plane in their world; once they identify left, right or behind, they know what to do without turning their head.

Shoreline creatures like otters, beavers, muskrats, manatees and seals develop small, soft ears that they can close when they go underwater. Our ears are definitely going smaller and softer, but closing them is another matter. Creating a thick coating of ear wax keeps the tender skin of the inner ear from drying out and cracking due to repeated immersions in water. A bit of wax is sufficient to handle several immersions per day. When humans go several weeks without water immersion, ear wax builds up. Despite the human urge to dig it out, the only safe method for removing large amounts of ear wax is with water irrigation. Not a single plains or land mammal suffers from excess ear wax buildup whose ideal cleansing method is forcing water into the ears. Ear wax is not a development that helps us with savannah living; it is an adaptation to frequent water immersion.

North American River Otter standing up on her aligned hips.

Talking about getting wet, no other savannah or plains animal requires regular bathing to remain healthy like we do. When humans don't rinse the oils off their skin for several weeks, we get skin irritations, sores, and smell awful. Other land animals aren't like that; a deer never needs soap and water. Someone might protest that they get wet in the rain. Yet a zebra, deer, dog or cat that secures a sheltered place and manages to stay out of the rain for eight months will still have no skin problems. Their skin is fine. They are not noticeably more smelly than a counterpart that got caught in a torrential downpour two weeks ago. Humans 'need' a good wetdown even when they aren't dirty. Humans need to have water on their faces regularly. We are healthier when we routinely wash off the body oils.

Then there is our nose. Most swimming mammals acquire muscles to pinch their nose closed underwater. That's easily accomplished with genetic 'more' commands when the nose already has some muscles. It's not an easy transition when nature has to start with an ape nose that consists of two open holes facing front.

The Proboscis monkey in Borneo is the best swimmer of all the monkeys. It gets its name from the fact that, like us, its nose developed a hood so the nostrils point down, not forward.

The proboscis is so adapted to a shoreline life that it evades land predators by leaping into the water, and water predators by coming on land. Its fingers and toes are slightly webbed. It swims in dog paddle fashion. Humans swim by reaching our arms over our head in the stroke or extending our hands to the side and propelling by pulling the arms towards the body. Both of these strokes make human swimmers faster than is possible by dog paddling.

While the monkey's nose and webbed fingers are adaptations to swimming, humans have more adaptations for swimming than the proboscis because we committed more strongly to a watery life and did so a few million years before them.

Proboscis monkey with a familiar nose

The nose change also changed the tenor of the Proboscis mating call. Male Proboscis developed an enlarged nose as a sexual attractant, the size entirely the result of sexual selection. While it may function fine in water, only the female has the nose shape mandated by a watery life.

The Japanese Macaques, also known as the snow monkey, are good swimmers too, yet spend much time on land. They eat insects, fruit, seeds, fungi and root plants, much like we eat apples, wheat, rice, mushrooms, potatoes and carrots.

Snow monkeys highlight another aspect that we'd expect: an increase in intelligence when a species is in transition from one environment to another. The snow monkey is the most north-living simian. Weighing between 18 and 25 pounds when fully grown, it's arguably the brightest simian species. Observers note that several food cleaning and gathering techniques and tool creation have appeared in the past 100 years which are passed on to other members only via training. Snow monkey groups have culture and customs they teach the young. The increased intelligence could have stemmed from adapting to a colder climate or to water; probably both played a part.

Japanese Macaques, aka Snow Monkeys. Babies and youth are almost flat-faced, with the mouth area protruding more by adulthood. While they can swim, they do not forage in water.

If a primate species came out of the trees a few million years ago, began living on shore and catching or collecting half of its food in the water, what changes could be expected as they adapted to that life? To swim efficiently, a straightening of the body would be useful. Legs would change from being bent at the knee to being aligned directly behind the torso to reduce drag and allow the feet to become flippers. In conjunction with that, the backbone would lose its forward curl and begin to permit a bit of backbending. This would allow the creature to swim with the entire body in a straight line just below the surface of the water.

The coarse ape hair would be a real impediment to fast swimming. In addition, hair would retain water, chilling the poor creature for a long time. This would be counter-productive when a cooling of the environment was the impetus to expand the habitat. Water creatures like the otter and beaver have thick, soft, furs with complex double density. Going from chimp-like hair to that kind of fur may have been too elaborate to manage quickly enough. Simply going bald was option two. It's not unprecedented; the manatee and the hippo took the bald route. They developed an insulating layer of fat below the skin, a layer we also have.

Alamy/Arco Images

A hamadryas baboon, a monkey adapting to savannah life, with hair and snout appropriate for that life. Its 'feet' are becoming simply knobs below the ankles and wrists.

To digress for a bit, why compare ourselves to the snow monkey and proboscis rather than the baboon? After all, the baboon's life today is almost identical to the long-held tale of our journey out of the trees. Baboons live on the savannah, organize into troops, and their diet includes fish, shellfish and small antelopes when it can get them. When the young reach six months old they spend most of their time with other juveniles, not their mother. Even the briefest look, however, into their life and biology makes the water monkey case even stronger. To start with the most obvious, the baboon's hair situation is the opposite of ours; body hair thickened while facial hair minimized, and in some of the five species of baboons, facial hair has disappeared. Their face didn't get flatter and the teeth get smaller; they developed a 'dogface,' meaning a long snout.

Face of the Olive Baboon. Has appropriate hair and snout for foraging and hunting on the savannah. When grown weighs between 30 and 90 lbs.

Their part-hunting diet gives them a large-opening mouth and sharp teeth. They walk on all-fours. Their young ride their mom's back like a pony once they are old enough to hold on well. They have no use for bipedal walking. Strictly speaking, they don't live in the savannah but along the edges of the savannah, since they require tall trees or rock ledges for sleeping at night; they do not sleep on the ground in the wild. They are intelligent, but as they develop into meat-eating predators they become more wolf-looking and acting in every way.

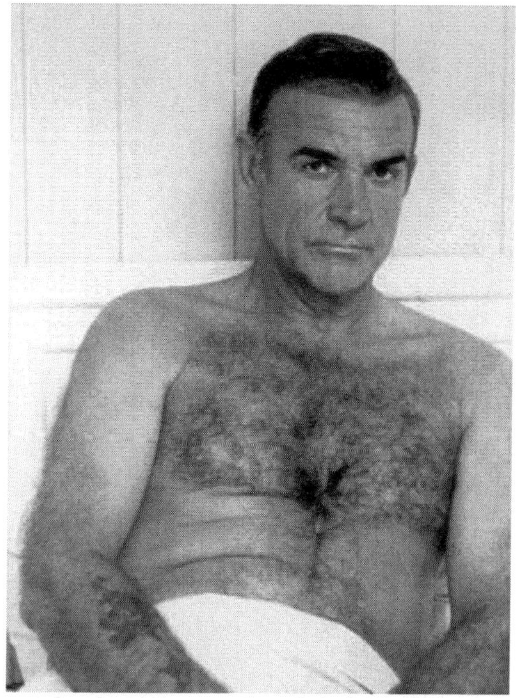

Human male, showing typical male hairiness. Sean Connery.

It's obvious from looking at people today that we didn't drop our hair in one spectacular gene mutation, but rather slowly worked it down via stages of 'less hair' genes. A wide variety of hairiness exists today. That oft-displayed generic naked man representing the human species is way off the mark; just ask anyone who has spent time in men's locker rooms. While our body hair has thinned out to the point where it doesn't provide much warmth (and easily dries out in a breeze), we still have plenty of body hair. Or rather, half of us do. Dimorphism in hair is a primary feature of humans. Males can have areas of hair so dense on face, chest, back and legs that it obscures skin color, while females don't.

Another human male with typical hair, Jack Black.

Just where you'd think a plains hunter would have the thickest patch of hair, on the shoulders to serve as sunscreen, humans have thinner, more spaced-out hair. When humans swim with their heads out of the water, the shoulders become their leading bullet-nose, so to speak; therefore, reducing the drag of springy hair on the shoulders makes sense. Interestingly, the direction body hair lays, meaning the easiest direction to push it down flat, varies between neck and pelvis. It may be mirroring the way water swirls around the body when a human dives in or is swimming fast.

Swimming and shoreline living permitted a more gradual transition from a mainly vegetarian diet to one that contained fish, crustaceans and birds. It also could provide the same nifty safety device that frogs and proboscis monkeys use: the ability to flee land predators by going into the water, and flee water predators by going on land.

Even though the hair thinned out and the bare skin underneath wasn't very robust in the sun, this new shoreline monkey could easily avoid midday sun without compromising food gathering very much. In time they may have covered their shoulders with animal skins. Even if they didn't hunt yet, there would be opportunities to secure skins from dead ones before the birds made a mess of them. The earliest known leather-tanning method is, no surprise, pounding the skins with rocks in the shallow water of a shoreline.

Third human male with typical body hair.

Going back to hands and feet with webbing at birth, this dominant trait is not reaching beyond our 40 million year monkey lineage to some ancient ancestor. It diminished and disappeared more recently, long after the Neanderthal split 400K years ago.

We wouldn't be the first land mammal to develop webbed hands and feet shortly after converting to a watery environment; we'd simply be joining the rather large crowd of current and past critters doing so. Our transformation from trees to shore-living took place over a leisurely three million years–some anthropologists believe the hip readjustment that was necessary to enable our legs to propel us during swimming started about six million years ago–so while our shoulders got bald, our legs got straight with our spines and we perfected the freestyle and the breaststroke, it would actually be an easier and less tricky genetic adaptation to have webbing between the fingers and toes.

We are born with an astounding amount of manual dexterity compared to other species. Why would our dexterity improve so much from our ape ancestry if for three million years we were just knapping stones, sharpening sticks and lying in wait for prey? The task to hunt big game and to eat that hunted food requires no more manual dexterity than a gorilla has.

Most anthropologists today connect early man with stone knapping to make sharp edges for cutting things back 1.5 million, and some, controversially, to three million years ago. Yet, when Bonobo monkeys are taught to stone knap with their more-curled fingers, the products of their efforts are similar to the earliest samples attributed to humanoids.[19] The closer the date of human-made stone products, the more controlled and accurate the strikes.

Is it plausible we have been using sharp edges of rocks for cutting as far back as two million years ago? Well, what do our teeth say? Most other creatures use their teeth for cutting. Our teeth say yes, our pearly whites haven't been either scissors or knife for a long time. Our ape cousins can cut leather with their teeth; we can't. In other words, while five

---

[19] Interestingly, in this experiment the Bonobo's motivation to learn how to knap stones was so they could cut *the rope* tying up a box of treats. It dangled where their hands could reach it but their sharp incisors could not. In another case, primate researchers at the University of São Paulo noticed that bearded capuchin monkeys knock rocks together, possibly to lick trace minerals off the newly-exposed surfaces. The flakes are indistinguishable from those attributed to early man, especially in those cases where both the shards and mother-rock are littering the ground together. It's likely monkeys have been doing this all along. Samir S. Patel, Archeology, Jan 2017, p. 21

million years ago a common ancestor had leather-cutting teeth, ours actually went duller and our jaw muscles became too weak to routinely tear through rabbit fur or deerskin. This means we survived for a very long time by eating softer things. Far from having a predator's chomp, our mouths could barely give a bad bruise on anything tougher than paper-thin human skin.

Manual dexterity in humans evolved slowly over three million years, and we see that progress in knapped tools becoming more sophisticated every 200K years or so. Our fingerbones were uncurling because when swimming, we held them perfectly flat on the return forward for another stroke. The use of the opposing thumb was improving.

Yet attributing stone knapping as the cause of the thumb improvement is like saying we evolved lips so we can drink from a cup without spilling. Rather, we invented cups because we *can* drink from them. Something one does well enough for *a million years*–drinking liquids, or stone knapping–isn't a reproductive selection driver. Put another way, to insist that the *only reason* our hands uncurled and dexterity improved was to bang two stones together for a few minutes per week is... not plausible. So what kind of activity would make minute improvements of manual dexterity become important and improve one's reproductive success?

Rope-making.

Survival that is dependent upon the ability to braid ropes and work with fine strings of materials to make light and flexible fishing nets would reward those born with better manual dexterity. Fishing with nets is the most efficient way to fish. Perhaps we could be called 'the species that rope built.' Rope-making would reward those who could 'stick to their knitting' for long periods of time.

Making a good net requires concentration, a tolerance for boredom, and manual dexterity. It requires lengthy training and attention to detail. These are not go-along traits with intelligence. These are entirely different traits from simply being intelligent. They are traits that chimpanzees and elephants don't have but human beings have in spades.

Farming, learning to read and write, manufacturing, and assembly lines; none of these would be possible if we had no innate ability to stick to an indirect and boring task with a payback long in the future. By 'indirect' I mean one or more series of steps that has only the possibility of leading to something directly resulting in food. Waiting along a deer path for a deer to come along is direct; spending the better part of a day manufacturing a rope that can later be fashioned into a net that next week will catch fish is indirect.

Once you start applying the theory of humans the rope-makers to the found artifacts, they make more sense. How would they carry ropes or nets? Over the shoulders, around the neck. It explains why the bone record often indicates early men and women were frequently carrying heavy weights for long distances. We find knapped tools but no storage containers because they made rope baskets and nets. Knapped tools would be necessary to cut leather or bark strips to make ropes, nets, and coverings since our teeth were not up to the task.

To those who insist this can't be true, then how did the gathering part of 'hunting-gathering' work, especially after fire was employed to cook food, back when people had no clothes or string bags? Everyone filled up a hand and then returned to camp? When monkeys forage, food goes from hand to mouth. By the time our ancestors were bringing food back to camp to share and cook, they were able to fashion baskets, nets or sacks. Whatever date is pinned on heating food at the fire, at least one year before that is when we invented the carry-bag.

Early man as rope-makers and net fishermen explains almost everything we find. It explains how the Neanderthals could have traveled so far and wide, confident of finding food along the way. If there was a body of water or river, with a net they could collect enough food to eat in two hours of work without even knowing the lay of the land.

Some anthropologists put forth a compelling case that when man began hunting large game, they did so in ways we characterize as 'cheating' today. They strung nets and rope between trees, then drove animals into it–netting land animals just as they had done with fish for the past few million years. Killing was done at close range with spears and clubs. They may have used natural formations to good effect also. They could drive animals into a muddy bank where they would become stuck, or scare them in the direction of a cliff so they'd jump off and be too injured to escape. Two people lassoing an injured large animal and staking their ropes to the ground or looped around a tree in opposite directions would make going in for the kill much safer. It would also compensate for the weak strength and light weight of a human, who couldn't possibly hold the rope freehand while a wild horse or wildebeest thrashed about.

Until stone arrowheads showed up, however, they did not hunt like we do today, stalking and taking aim. Spears, with their large stone tips, were excellent for driving into an animal tied down or trapped, but the stone was too large to rip a hole that big through an animal hide when thrown even eighty feet.

Early spears could have initially been used for hunting of large fish and water creatures. Fishing spears wouldn't need stone tips to work well; bone tips would suffice. This could explain why stone points small enough to be affixed to arrows did not appear until 50,000 years ago in southern Africa.[20] A small stone tip would be needed to pierce the thick hides of large game.

---

[20] The location of this earliest arrow find, far from the Mediterranean, plus the date indicate that humans who headed north around 60K did not have bows and arrows. It appears they were reinvented in several locations in the world more recently.

When most of Africa was plunged into a 30,000 year long drought during the second-from-last glacial period which ended about 60K years ago, our toolmaking, ropes, nets and control of fire plus a few million years of a fish diet put hunting instead of scavenging large game in the range of possibility.

Scavenging was an option all along, just as it is for ape species today. Finding proof that humans carried big game parts to a safe spot and cut meat off with stone scrapers does not indicate they hunted that game. It does mean they were already compensating for inadequate tooth sharpness by using knapped stone tools to cut their food, strip bark to make rope, and for defense.

When we find ancient remains that shed light on early man's diet in Africa before they migrated north, this is exactly what we find; a small portion is big game. The heaps of large bones seems huge until one does the math. Considering that small bones would deteriorate completely, what looks like 30% several thousand years later could have been more like 5% at the time, with most of the frail fish bones either not even thrown on the bone pile or totally disintegrated. Finding 600 bones left over 200 years is only three per year. But usually the number is more like 600 bones over a thousand years.

Recently, tentative evidence has been found of boat-building tens of thousands of years earlier than previously believed. If it is correct, then boat building may have begun 600,000 years ago by the common ancestor of both us and the Neanderthals. Like most advances in the days before writing, boat-making was probably lost, then rediscovered, several times before becoming a species-wide custom.

The Greek island of Crete, too far from the mainland to get to by swimming there and the sea too deep to ever have had a land bridge, has signs of Neanderthal occupation 110,000 years ago. Much later, humans became good enough at boat building to make it to Australia an estimated 50,000 years ago. Their boats were seafaring enough to hopscotch from mainland Malaysia and several islands carrying women and children. It wouldn't be a stretch to imagine the trip was made not once but several times.

Why did we lose our finger and toe webbing? The reason had something to do with that glacial period. Whether there were practical, hunting-related reasons that made it advantageous or opposite sex preference, we may never know. Don't scoff at opposite sex preference: it created peacock feathers and size dimorphism between males and females.

Another possibility is that Neanderthals still had webbed fingers while African humans, having gone more inland to live, lost their webbing. When our ancestors left Africa 60,000 years ago and arrived in the European and Asian habitats already occupied, one of the most profound differences may have been webbed fingers.

On the other hand, it is just as likely that Neanderthals lost their webbing long ago–too prone to get frostbite– while early humans ventured north with theirs intact. That scenario would be harder to prove since today's native Africans have no Neanderthal blood and also do not have webbing. Perhaps both of them lost the webbing for entirely different reasons. All of this is pure speculation.

While the theory may be speculation, the webbing is not. It simply crops up too often, is a dominant trait, and is too obviously the type of webbing that would assist with swimming for it to be 50 million years in our past instead of 80,000 years.

If our African homo sapiens branch was a water monkey for three million years with possibly only the last 70,000 to 200,000 shifting towards life as a big game hunter, then our digestive system would also tell that story. Does it?

Perfect Food

Every food is a perfect fit for some creature. Each VS finds a niche of edible food and in time it becomes the perfect food for them. Their digestive system adapts so well that within the limitations of their chosen diet–doesn't matter what it is-- it becomes *exactly and only* what they need. Two creatures diverge millions of years ago and today one is a deer who could eat twigs all year and thrive, and another is a cat who needs only meat. Neither suffers from poor nutrition when they eat the food to which they've become adapted. Both would suffer indigestion if forced to eat the food of the other. As the two lines diverged, each experienced a gradual shift in diet. Earth has a rich history of herbivores becoming carnivores in the course of a few million years. Birds, for example; they are evenly split between predators and prey. Some are herbivores, some are carnivores, and some are both. A person can't tell at first glance which is which. The birds that feed at the birdfeeder are herbivores, but the carnivores–robins, woodpeckers, crows, cliff swallows and the like–are absent.

Birds are proof that appearance may change very little while diet does a complete flip. Bird's feet, feathers, head shape and wings can be virtually identical even when one species eats only insects and protein and another exclusively seeds and berries. Then a third may eat both. Only tiny adaptations in the beak might give a clue. Still, because even beak analysis isn't enough, only observation or someone telling you what the bird eats will inform you on their diet. Similarly, only our teeth and digestion system will render a clue to what our diet was 300,000 years ago, not walking or using hands to pick things up or being hairless.

We know that humans used to be herbivores. We also know we are strongly omnivore now. We call ourselves predators, but that is a newish designation; in reality our digestive systems are in transition, and like all transitions involving a widespread population it progresses in a choppy fashion, affecting some pockets of the population more than others.

An interesting sidelight is that most creatures have things they enjoy eating, things they will eat occasionally, and even things they will eat but aren't that good for it. When a dog eats a chocolate chip cookie it gets some calories from it, but it's not something a dog can eat every day and live a long life.

Many species, not just our own, have a distinct preference for a delicacy. A delicacy is defined as something scarce, hard to find in the environment and often not particularly good for the creature. The species definitely could not subsist on only the delicacy, although it will go to great lengths to get it. Instinct has imbued such a desire for it that if it were possible, the creature would eat it to excess and damage its health.

In past eons, this very seldom happened. It wasn't a worry that say, a bear, would eat only honey, or a cat would eat only thigh meat from a cow. A bear needs a great deal of nutrients not available in honey, and the cat needs the full range of bone, blood and other aspects it gets from eating the entire mouse or mole. When you see bone and intestine parts listed in cat food, they are not ripping you off by adding scrap parts into the food, they are balancing your cat's diet to its long-adapted operating system. If they don't do that, your cat might live only 15 years instead of 20. A cat is adapted to tear the skin off and then eat the whole creature, bones, intestines and all. This is what it has done for the past million years and therefore constitutes the diet it must have to remain healthy.

Humans also have delicacies, and you know what they are. They're the things we love to eat but are bad for us in large quantities. Like the bear, we were never meant to have honey every day. As much as bears like honey, not a single bear became a beekeeper in the past 20 million years. No matter how much it loves beef, a cat will never keep a herd of cows.

Conversely, instinct isn't motivated to make our best suited and most common food the most appealing to our taste buds. It assumes we'll eat it the most because it's around, and our lifestyle and environment makes it common. We don't hate it, we just don't feel strongly about it. We'll eat it to quell hunger. When we're really hungry, it tastes wonderful.

Back to cats again, someone did a study on cat's preferred food. The researcher performed his study on feral domestic cats accustomed to the native environment.[21] In the wild, cats hunt alone, eating mainly small mammals, sometimes reptiles or amphibians, sometimes sneaking off with the young of bigger animals. They'll eat birds and eggs, if they can get them. The largest percentage of their diet is mice and rats. However, when given equal access to 1) a mouse or to 2) a piece of chicken, beef, pork or fish, the test cats always picked the non-mouse choice.

While written in quite scientific terms, a little of the researcher's astonishment crept in, because in over 200 tests, *not once* did one of the cats select the mouse. Zero times. Despite the fact that the normal diet of these wild cats was 80% mice, rats, moles and birds, they had no preference for them. They had a preference for foods that one could assume they almost never came across.

Humans are like that too. There are foods we love, and foods that are best for us. The best-for-us foods will not be things we crave but things we find tolerable. Not delicious. Or we need to make them delicious with spices and sauces because they simply aren't very tasty otherwise. Because a stomach and digestive system takes a long time to turn (oh, a pun), our best-case diet will be amazingly similar to that of many generations ago. What this also means is we remain the

---

[21] I think the study took place in India, but it was published many years ago and I could not find it on the internet.

healthiest with foods we historically dined upon for the past million years. Food we adopted only 200,000 years ago will get processed a little less well, and food we adopted only 10,000 years ago will disagree with us in the next few hours.

What will a nutritionist recommend as the ideal, digestible human diet? It's not even difficult; the consensus is: eat fish every day; roots like sweet potatoes, taro, turnips and carrots; seeds like rice and almonds; and fruits like apples, figs and grapes. Beef, meh. Pork, meh. sugar, meh.

Our own health repercussions tell us that large game is a recent addition to our diet. We still have trouble digesting it. Humans are the only mammals which get gout. Gout is an inflammation caused by imperfect digestion of red meat and other recent dietary additions. Humans have only a small amount of the enzyme called Uricase, which breaks down a part of red meat into a harmless, easily peed-out compound. When a person eats red meat almost daily and is on the low end of the uricase range, leftover bad chemicals accumulate in the lowest part of the body, the foot.

Why do we have so little? One reason is because digesting fish doesn't require uricase. Rabbits and mice have four times more uricase than we do, and they aren't even predators. We have some uricase, with the quantity varying by person. It's a clear sign that red-meat-eating a few times per week is new to us and we're still adapting, suggesting a less than 100K timeframe.

Nutritionists simply state outright that it's fish, not beef, we should be eating because it gives us a 'better' protein. Fish is not inherently a better protein to all creatures, but it is to the ones long adapted to eating fish.

There are two caveats to flat out declaring all humans have one perfect diet: one, people whose ancestors lived for many generations in very cold climates like Eskimos and Laplanders have adapted to a diet consisting only of fish for several months of the year. It stands to reason that switching to a diet that's even 50% fruits and veggies in one generation will not sit well with their digestive tract. Two, when populations are homogenous for dozens of generations, the diet they have becomes the optimal one; yet today there are few such homogenous populations left. When a Swede marries an Italian, or a Hawaiian marries a German, what gut do their children inherit? What diet will work best? It stands to reason those offspring will have to pay close attention to how they feel after eating to figure it out.

When did we start using fire? Evidence gets understandably vague the older a site is. Currently, sites that indicate early man had a fireplace going around 400,000 years ago are compelling enough to be accepted. Claims of earlier proximity of man and fire are hazy. Dr. Francesco Berna's research at Wonderwerk Cave in South Africa provides evidence of actual fires, not blown-in ash, over one million years old. This cave used to be near the shore of a lake. Near the front of the cave, chipped tools were found, ranging from the Oldowan[22] style [2 million years ago] to better-looking tools made 800,000 or less years ago. In this very stable cave which has been there for millions of years, stone tools were found that are not made of stone from the cave but carried in. Current thought is that early humans harvested fire from lightning strikes and kept it going, but may not have been able to generate it from scratch from flint or rubbing two pieces of wood together until much more recently.

Keeping a fire would have been done, lost, then captured again thousands of times over tens of thousands of years before even 80% of small tribes had a fire going at any one time. It's plausible that the fire going out overnight was a primary reason to maintain good relations with the neighboring tribes, so someone could hike out with kindling to 'borrow' some.

---

[22] Early humans made clumsy lumps of chipped stone, with no improvement, for a million years. We call this the Oldowan style.

Two caves in Western France with signs of Neanderthal occupation have thick layers of fireplace ash, indicating the cave was used to stash fire for many generations. For just as long, they chipped away at stones for hand tools while sitting around this fire. The current theory is that early man remade tools every week or as needed, not hanging onto any of them.

All the sites where early man the shoreline dweller had a firepit that was used daily for generations are long gone, ruined by rivers changing course, lakes changing shape, and most importantly, ground up by incoming and receding glaciers and their after-peak water runoffs. In 1.5 million years mountains have worn down and other land has risen due to weathering and plate tectonics. Similar to finding humanoid bones, human-built firepits would be preserved only when the early humans were far from home, going through inhospitable, even barren land. Lush waterfronts would scarcely retain a molecule of a firepit abandoned 500 years ago, much less 400,000.

Beyond the warmth, light, and discouraging of nocturnal predators, the odds are good that the primary impetus to capture some fire and keep it going was to cook food. A fire softens foods that are hard or tough, something easily discovered when foraging in burned forests after lightning fires. In some cases cooking a food removes toxins, making the food more edible. It will never be possible to tell whether the warmth, protection or cooking aspect was the most compelling reason for our mastery of fire. What we do know is that our teeth should look more like the otter's and the baboon's teeth, sharp and pointy in front for cutting off bite-size bits, if we ate our fish, tuberous roots and scavenged game raw up until recently.

Root food like potatoes, carrots, turnips, yams and taro, and tough nuts as well as tough meat are made much softer and in most cases tastier by cooking. I'm not going out on a limb when I say both human and Neanderthal teeth have not been sharp enough nor did our mouths open big enough for teeth to be our primary method of attack or defense in the past million years. It's plausible we have been cutting our food with stone implements for two million years and cooking it when we had fire for a million years. Not having sharp teeth has not been an impediment to reproduction or feeding the young for a long, long time.

Older members who lost many teeth and those with tooth problems could eat well. Therefore it stands to reason that the older members of the group were the most interested in keeping the fire going. Since we were already a species capable of learning and had young with a long childhood, improving the diet of the elderly increased their lifespans and therefore their ability to watch over the young.

A fire is hard to move and even harder to restart, so once a group commits to having a fire they are motivated to stay in that area for at least a few weeks. Or months. The more early man depended upon fire, the less they wandered. Someone staying by the fire was necessary. Realistically, it was usually an elderly or recuperating injured person. These same people could watch over the young. Reduced childcare demands enabled reproductive-aged parents to devote solid blocks of time to gathering food.

Like most things involving nature, it's all chicken and egg; things happening so concurrently that one can't be tickled out to be proven to cause the other. Grandparents watching and teaching children was probably concurrent with fire and cooking. It's hard to grasp how demanding a hand-to-mouth existence is, and how incompatible it is with carrying around a baby for two years and keeping constant watch on several young at a time for another six. Adults of child-bearing age were the only tribe members capable of doing the hard work of obtaining food for the group. Long childhoods jeopardized humanity's survival since the quantity of young under ten would nearly always outnumber adults. Attrition was high.[23] The trend towards grandparents and injured group members raising and guiding the whole group's young was the lynchpin to making the long childhood work.

Our young have a strange dynamic not seen in any other species; when kids are combined into a group of several children, they are more docile and obedient than when solo. Parents are hard-pressed to get one or two kids to be well-behaved for several hours on a Sunday, yet a teacher can control 30 kids all day with less effort. We're so used to it that we no longer perceive how spectacularly unusual that is. If one or two is a handful, eight or twenty should be simply impossible. But the opposite happens. It's easier to keep 30 kids quiet, sitting down and doing a task than two kids.

Instinct within the child brain makes them feel safe when other children are present. How many times have I observed a crying baby in a stroller who stops upon spotting another baby in a stroller, and the two of them lock eyes for as long as they remain in sight? Even if they are not the same age, it's transparent that the younger one feels more secure upon spotting another child in its same predicament. It is as if children are hard-wired to feel edgy and suspicious of motives when only adults are around. Being with other children is what he or she really wants to do.

One explanation for the 'group effect' of young humans is because that is how they survived. Compare typical parents and then older people minding children. When a child misbehaves, the young parents leap up and approach the child to swat or pull it away or physically alter the behavior. An older person with the same provocation will say 'Don't make me get up!' or some similar admonishment. Nine times

---

[23] Even as recently as 1900, three out of every ten babies born died before reaching one year old. Before a few hundred years ago a newborn baby had less than a 40% chance of making it to 16. There's no reason to assume early man had better statistics. Information courtesy of http://www.pbs.org/fmc/timeline/dmortality.htm.

out of ten, the child does not make him or her get up. They modify their behavior.

Approaching and physically correcting a child doesn't even seem to be necessary. Children with wheelchair-bound parents show the same range of obedience as children of parents who leap up and modify behavior with physical contact.

It is curious that leap-up-and-lay-on-hands parents have to do this frequently with small children, but day care providers and kindergarten teachers can keep better control of five to ten times more children with just words and facial expressions. The only way this could be hard-wired in children is if being watched over by the less-spritely members of the group, with Mom and Dad nowhere in sight, has been going on for hundreds of thousands of years, not just the last 8,000. Children who did not have that instinct wandered off or disobeyed, usually to their peril. A lack of 'obedience in groups' instinct would significantly affect reproduction since that child wasn't going to make it to 16.

Children always outnumbered the older or handicapped caretakers by quite a bit. Because only one adult out of four or five would even have a surviving parent, obeying only a relative is not part of the metric. In ape species today there is some shared parenting and youngsters do play together, but for the most part the co-parenter is a relative and the young do not have that compelling drive to hang with other youngsters.

This also explains the growing amount of bone evidence that people who lost legs or arms sometimes lived for a few years after the event. Why would their group chose to provide food for that person when life was so hard? Because a firetender was needed. Because children needed to be watched. Because having an adult do those things was a net gain for the group.

Did our ancestors live in caves? We find evidence of Neanderthals and humans eating lunch, maintaining fires, drawing pictures and littering in caves. We've found bones, but couldn't be sure they weren't dragged in by a cat. Or bear. Plenty of caves that were around 25,000 years ago have no sign of either humanoid ever visiting. No remains, no evidence–which means they may or may not have been there. To have everything over 10K years old weathering or rotting away is the norm, not the exception.

Caves are not that common. Lush environments with only minor hilliness will have few caves in dozens of miles. Areas where caves are found are often not a good place to live. They can be miles from the nearest river or body of water. Humans may have lived in caves like birds will nest in roof eaves; will do it if there is one, but not a big part of their survival strategy. Stuff in caves is all we're going to find, however, so people will continue to say all early men lived in caves. That is like saying all things lost at night are under the lamppost, because that is the only place we find them.

In large swaths of Africa and South America, naked monkeys and apes live today with little more than impermanent bedding in the trees. Deciding our ancestors needed 'houses' 200,000 years ago and trying to find them says more about what we like today, not what our ancestors needed to live. As far as we know, covered bedding areas for protection from rain and snow were built before 100,000 years ago, but housing, envisioned today as a place to live and keep one's belongings, began less than 14,000 years ago.

At the Terra Amata site in France, there is a museum containing evidence that Neanderthals were making impermanent shelters with collected stones and sticks. It was optimistically dated by DeLomey, the digger, to 380K, but more realistically dated to 230K by later archaeological review. The shelter, 8 x 4 meters or about 24 x 12 feet, was located on a beach and had been occupied off and on for thousands of years, accounting for dating vagaries.

Terra Amata sleeping hovel our ancestors built, 400K to 20K years ago. Current theory is that when either humans or Neanderthals were not in caves, this is what they felt like building. Only dating techniques will tell the age because there's almost no change for hundreds of thousands of years. Size: 12 x 24 ft. 4 x 8 meters. Sleeps 30.

If someone 120K years ago dug a hole to store something or to make a mound to sit on, whatever fell into the depression would look 200K years older. All sites with multiple habitation over thousands of years have inevitable difficulties with proper dating.

The pros know that evidence from a sole habitation site is twenty times stronger than 'beliefs' of age produced from layers of habitation. Headline writers do not know this; more often than not they lump everything found on the site to the oldest date. Whenever someone digs a posthole, who knows when it was dug. We can only best guess. Half the time it puts older stuff into newer layers–accounting for archaeologist shock when an artifact thought to be 15,000 years old ends up being 35,000 years old.

Several other locations in Europe have found the Neanderthals pushed stones together in circles or ovals to brace sticks that then had other sticks or skins flopped over the top to create shelter. What's notable is that housing didn't improve in over 300,000 years; the same shove stones, collect sticks, and drape with hides could either be 380K or 50K years ago. Only dating techniques will tell us the age of the site. Anthropologists estimate the amount of people using the hut by how many could cram in to sleep. No one disputes this shelter served a large family group that used it only in bad weather.

Using hides for the top of the shelter and for bedding are not, however, a sign of hunting; they could be more readily acquired through scavenging. Several sites in Europe have been found with lots of large-creature bones nearby. At one site they even used mammoth horns as the supports for the draped hides.[24] It should go without saying that collecting the horn of every mammoth that died of any cause in the past 100 years within a ten mile radius would be simply a matter of having people agree to carry it.

Prehistoric sites containing lots of big-game bones are taken to 'prove' early man was hunting. Not necessarily. One, big-game kills are where it's most likely that the original predator(s) cannot possibly finish it all. Hauling off a leg or cutting off swaths by the ribs is what scavenging early humans would do. Later , we'd go back for the horns or even the skull, to make bowls or other tools. If early humans were hunting, most of the time they'd catch less dangerous prey like deer

---

[24] I've long quipped that the pre-historic fat woman clay figurines weren't religious, they were used to hold the door curtain open or closed, and the shape of it reminded everyone which menopausal lady they'd have to answer to if they messed with it. While my intent was to be rudely flippant, it's now turning out that these figurines are found near openings that likely had a door flap.

and hedgehogs. Dig reports don't mention more hedgehog bones than mastodon in the bone piles.

Two, the quantities sound impressive, but several thousand years is a long time; some of the sites have deposits ranging over 200K years. Finding 'hundreds' of bones is not much over that timespan. Those hundreds are more in line with a single bone added every three years, not once a month. Once a month would be the quantity if the Neanderthals had a diet of even 10% big game. In fact, if getting a piece of mammoth or rhinoceros were a big deal, it goes far to explain why they hung onto the big bones and why there isn't a single full–no, not even half of a skeleton.

Don't even try looking up the reasons why Neanderthals are no longer with us. Someone, at some time, has used every single human characteristic as the reason.

For instance, one can find articles making a compelling case for Neanderthals having a diet of mostly big game while the newcomers had a broader range of food including shellfish, fish and a variety of veggies, and that broader range of diet was why they won and Neanderthals lost the survival game. Then, another author makes just as strong a case that the Neanderthals were slow, inept hunters while the newcomers were skilled hunters with superior weaponry, which is why they won.

Bluntly, people make stuff up out of prejudice. Both of those can't be true. Other people, who should be better at math, publish with a façade of authority that there were few Neanderthals and they didn't thrive, despite having a range from England to Korea, from Finland to Turkey, and from coasts to mountaintops in just 350K years. Calculate a 20-person family group having 120 sq. miles–even though 40 miles is more plausible-that puts 22K in today's Germany and 41K in France, which is not even 5% of their range.

Current evidence shows Neanderthals may have invented boats and the stick-shelter-with-indoor-fire design first, but the newcomer humans learned how to do both quickly once they ventured North. Now the evidence is adding up–and this could change–that the Neanderthals were water-oriented boat-fishermen when possible, building craft over 100,000 years ago that could hold several people and a decent amount of supplies. It's obvious early humans had no boat technology (not counting riding on logs) prior to meeting the Neanderthals since similarly old human habitation on islands off the coast of Africa hasn't been found. On islands off the coast of Africa, even those very close, human habitation is newer than 10,000 years ago.[25]

The Neanderthal physical adaptations–shorter legs and arms, stocky build–were adaptations to cold and tell us nothing about diet. One thing we do know from the bone evidence is that humans were better at running than Neanderthals. The sites and artifacts point towards Neanderthals staying close to the water once they left Africa 400K years ago and so they were better swimmers than they were walkers. Due to drought, however, the concurrent humans in Africa had a 50K year jump on them in adapting to long walks over grassy terrain.

Modern humans were able to hunt medium-sized game, if not large game, when they left Africa 60,000 years ago. For discussion's sake, say they spread at a pace of one hundred miles every three generations. But where did they go and when? Over that span of time both Neanderthals and humans occupied certain sites.

Until recently, everything found that was a bit advanced–jewelry, musical instruments, artwork, tools–was automatically assigned to the humans. Even today, if it's newer than 60,000 years ago it is attributed to humans until

---

[25] Pemba island, 20 miles off the coast of Africa, is typical. The first humans came in 600 AD. No one boated over to Madagascar, about 200 miles off the coast, until about 100 BC. Sherbro Island, less than two miles from shore, was first inhabited about 500 BC.

proven via DNA or protein-testing that it's Neanderthal–and lately testing is proving just that.

So our map of early human migration is being redrawn; the artifacts used as proof of human habitation now show humans weren't there. An unfortunate past prejudice that these artifacts looked 'too nice' to be Neanderthal is creating an embarrassed blush over the anthropology community.[26]

Humans, having bows and arrows by 30K years ago, were the ones that hunted the mastodon and several other large creatures out of existence in Europe, and did the same after going to the Americas around 25K years ago.

Neither our stomachs nor the found evidence, however, indicate that 300K years ago we'd transitioned to a hunting creature.

Groups of early man, with no useful teeth, nails or speed, would most likely tune in to kills going on, let those predators eat a bit, then storm in with their spears–best used with thrusts, not throwing them–to drive off the now-full killer beast, then hack off leather, horn and meat and high-tail it back home.

Assuming the same learning curve based on how fast their housing design improved, which is zip, it's likely their large-game gathering prowess also stayed like this for a very long time. To throw in one more odd thought, perhaps man didn't domesticate wolves into dogs so much as follow them around to get some of their kill. Maybe it was rushing in to help when it looked like the prey might get away that was the beginning of a close friendship. Maybe the ancestors of today's dogs fed us as pets for a long time, and only recently did we return the favor.

The truth is, for two million years our ancestors did about the same thing generation after generation, with dinky

[26] If you want to see a typical example of the 'hey it's too nice so humans must've been here' hubris, visit http://www.bbc.com/news/science-environment-18196349

improvements in stone tools taking 200,000 years to manifest. These changes were more likely a factor of improved hand coordination making them possible, not advancing intelligence.

Then 60,000 years ago the African humans developed a wanderlust that is unparalleled in the animal or plant world. It was frog night for humans. Within a mere 15,000 years they were everywhere: leaving Africa, family groups made it to Norway, Korea, Australia, China, India and Ireland. Arguably, there is not a single other species whose range increased like that in 15,000 years as a species-originated wave. Species like frogs are all over, but it took so long that they've become completely different-looking, with different features and strengths and diets. No one type has a range of more than several hundred miles even after millions of years.

What caused the wanderlust? It wasn't because the parents wanted their adult children to move away. Like a released spring, they not only went fast but kept walking, passing up perfectly fine streams and lakes in their progress.

Then the door closed again. The glaciers came back. The ice sheet peaked 25,000 years ago, when ice covered much of Northern Europe and Asia as well as North America. During the thousands of years around each peak ice coverage, Africa would become arid, with rainfall diminishing by 90%. It may have been drought that elbowed humans out of Africa 400K years ago and again 60K years ago, but that still doesn't explain how they got to Korea, Germany and Australia so fast that last time. Glaciers were receding and the climate began warming around 14,000 years ago, and the Earth settled into an interglacial period, with weather more or less as it is today, around 11,000 years ago.

Evidence our human ancestors hunted big game or had cooking fires prior to 600,000 years ago is generally open to other obvious interpretations. Holes in hip bones held to be evidence of spear-hunting could have been driven in with a stick after death so two people could carry the hanging scavenged meat back to the main group. If a heavy portion of an animal was going to be carried five miles, driving a stick through the hipbone instead of the meaty parts would be common practice. It's more realistic that a nice round hole was created by driving down into a hipbone while the bone was laying inert on the ground than from a spear thrown into a live fleeing animal; the math on the kind of force needed, both weight and speed, doesn't mesh with a freehand throw from a 100 pound early human.

Stone shards and cutting tools were not necessarily for hunting and meat-eating; a more common use would be for cutting up roots and vegetables into bite-size bits. Hammer stones would be used to open shells. Small sharp slivers of bone and stone would be used for digging out nutmeats and edible bits from crustaceans. They would also have been used to make strips of fibrous bark to make rope and bindings. Bindings could be used to make shelters.

With bindings and rope, early man could make nets, snares, bags and baskets. They could bind up more items to carry along than only their hands would hold. Using a stick resting on the shoulders of two people, they could carry goods and even people long distances while having their hands and bodies free in half a second if the need arose.

When you are a bald, weak creature, having ropes and nets means eating well and easily making shelter. We are a species that loves to move around. Up until the past few generations there were whole populations in North America, South America and Australia that moved seasonally every year. But why talk about long ago? To this day when a family has a bit more money than necessary for a comfortable life it is common for the next purchase to be a second home. This holds true regardless of country or culture. One could say we are hard-wired to spend time in a second home. Even our word 'vacation' meaning time off from work carries the baggage of also meaning leaving one's house and going to live somewhere else temporarily.

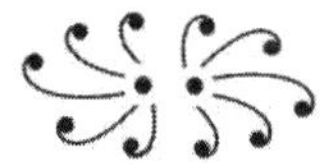

Why does anyone say early man came down from the trees to hunt in the plains three million years ago? If our direct ancestor is so easily proven to be the water monkey, why do some foot-stompingly insist on an unsupportable fantasy? Well, two reasons: one, the locales where they find early man bones are nearly always barren areas today. Bones of creatures that died where they lived, on shores and areas teeming with life, would never be preserved. The only ones that had even the slightest chance of being preserved would have been far from home. They would have had to die where predators could not get to them. Or die just before an unusual climate event that somehow prevented any predators, mold or vermin from taking their fill.

Scientists are not allowed to use the Judge Judy rule but must base all opinions on the found data. If they found humanoid remains on top of a mountain or in a cave, then that is where they must have lived. To state 'they were only visiting' or 'they got lost' or 'their relatives kicked them out so they wandered off' is simply not done. They found the bones there so they lived there. Period. Never mind that at the time it was an inhospitable place. Never mind that in the history of time no mammal just parks itself next to the dead carcasses of its companions and keeps on living there. The Judge Judy rule says that if you find human remains in a cave, you can bet your booty humans DID NOT live anywhere near there. They lived a half mile away. Or ten miles away. But not here.

Two, finding the earliest-known evidence of anything involving our ancestors is a big deal, and many artifacts are anointed with the earliest possible date, even when the most likely date is thousands of years newer. Many of these early dates are simply the optimism of the discoverer. They self-correct with a little more scrutiny and a couple of tries at carbon dating or other scientific dating tools. But does that always happen? No, it doesn't. I take the stated age of new discoveries with a grain of salt until there is a year or two of peer review. I read the actual details of what was found, not only the interpretation of the finder. Often it is only the headline writer being overdramatic.

That's how hunting got pushed back to three million years ago; stone flakes in the vicinity of old bones that have a little groove that looks cut with a sharp edge. But we already know that our ape cousins scavenge meat off bones, and we know our ancestors were chipping off sharp edges for cutting for *more than a million years* before someone thought to bind it to a stick. So long did our ancestors substitute sharp stones for incisors, in fact, that our teeth got wimpy and rounded, which evidently made no difference to anyone's reproductive prowess.

Sometimes I suspect I throw around the phrase 'a million years' so much that meaning blots out. Imagine a species did something that wasn't very helpful to its survival since the time of the early Egyptians to today. That would be 6,000 years. Pretty unlikely a maladaptive behavior could last that long in any environment. Now multiply it 167 times.

Imagine a species did something exactly like their parents did for 6,000 years, and then again for 167 repetitions. I'd call that not going anywhere. That's staying put. When early man uses stone cutting edges but doesn't tie them to sticks for a million years, that's not evolution. That's a long time of being 'finished' and then, in a time span likely to be less than 5,000 years, advancing. They weren't slowly advancing all along. Look at apes today, using a straw to poke into termite nests and then eat the termites that cling to it. Their ancestor ape probably did that same thing 300,000 years ago. They're not evolving into a smart species. That tool-using doesn't have to be headed anywhere. It's good just as it is.

Go Take a Warm Bath if You Disagree

While you soak, ponder how many savannah creatures like wildebeests, bison, zebras, baboons, giraffes and ostriches do clean-off-the-dirt bathing. Rolling in the mud or dust doesn't count. That's the opposite of a human-style bath. If you don't think so, wear mud on your shoulders for two months and tell me if your skin is fine with that.

We are the water monkey. We are a creature who thrives on daily bathing, who has lost the thick hair that once covered the body, a creature for whom long hours in the sun is not good, a species that lives along rivers and lakes because we're instinctively drawn to them, and the foods that keep us the healthiest are fish and fruits and roots. Our feet have irrelevant toenails and are soft and wide like paddles, not hard with points or thick claws to kick off from packed dirt quickly like savannah animals. We like to cook our food, something not terribly risky or attention attracting on the lush bank of a river but difficult to contain in the middle of a dry field of grass, not to mention visible for miles around, attracting all the wrong sorts to our dinner.

The early man dioramas in museums? Imagine a spot on a lush riverbank, a lean-to of sticks tied with rope to a low branch off to one side, with Mom and her sister working on making a net using a stone blade to cut strands.

One Dad naps under the lean-to because he worked hard yesterday while another Dad stands knee deep in water watching while two pre-teen boys practice throwing a heavy net in the shallows. Grandpa, the 6-fingered firetender with an oddly-bent ankle, fries fish fillets speared on sticks over the fire and with another stick flips cut-up roots wrapped in leaves in the hot embers. A big kid plays some kind of stick game with several smaller kids about ten feet from the fire. Extra food wrapped in a few hide sacks dangles from a high tree branch a short distance from camp. The ratio of kids to adults over 12 is about 60:40. This is how it was for a million years or more. There is not a single sign of a savannah life in our bodies, our diet, our skin, our preferred places to live.

All I see is the water monkey.

The Stopgap Measure

Nature prefers instinct. When it must, it gives a creature a brain. Most creatures with brains will tax their brain to its very maximum several times per year. Even once a week. That doesn't cue nature to give them more brainpower. Those species carry on with about the same smarts for hundreds of thousands of years. Foxes, squirrels, crows and giraffes are getting neither smarter nor dumber. Within the species there is a range but 'quick thinking' never became a compelling reason for a selective sweep. The odds are good that most animals are as bright, on average, as they were 200K years ago.

I could be wrong, but I'll stick my neck out to say the only impetus to increase intelligence in a species is a drastic environmental change with a concurrent increase in complexity of finding food. Adapting merely to eat a plentiful food source, like grass, does not make anyone smarter. When the horse evolved from a smallish, claw-footed creature into a horse, or some kind of land creature evolved to a manatee, they did so without any bump in intelligence.

Most animals and birds stay in their niche, not changing from sky to land, or herbivore to predator, or from trees to water in a very long time. An 'increase intelligence' adaptation is not in their recent past. If they once had it, it was fifty million years ago, not twenty times closer, so is not readily available for occasional trying out again like webbed fingers or hair all over is with people.

Nature has no inclination to overendow in the brain department. Over time nature moves more of a species' daily and seasonal behaviors into instinct. This puts great problem-solving and instant response times into a very small space. It also means the youngest of the species is as good at complex things, like flying or evading predators, as the oldest of the species.

Instinct is like Tae Kwon Do. Initially the training combines motions into a set. Trainees learn the combinations. In time, if the first move is initiated, the others just follow out of habit. The motions make little sense while learning them. They're just punching and swinging in the air. A person can learn several routines in a room by themselves. Then the very first time they encounter a thief with a knife in the street, they may engage the set of moves that comes to mind and have them work to disable the knifeman. Even if the bad guy responds in a normal fashion to the first motion, grabbing that hand or deflecting it, it almost doesn't matter because the second and third motions come so fast. A person may never have disabled a knifeman before in their lives; may never even had hand-to-hand combat. Yet in their first encounter, speed and lack of hesitation plus attacking vulnerable spots on the opponent's body results in the desired outcome. No one has to explain to trainees why or how these are vulnerable spots, the neck or the kneecap or solar plexus. The motions simply go there.

Brains need practice to do what instinct does instantly. Puppies and kittens practice hunting with their littermates. Children play at whatever the adults are doing. Playfulness in any species is usually practice and how much a species plays may be a good way to measure relative intelligence.

Nature prefers small brains and big instinct. If the digestive tract on Winter Moths and the eyes of salamanders deep in caves can disappear from lack of use, would nature ever get rid of a brain? Yes.

Take the sea squirt. While many species have a larval stage which then segues into a far more complex creature, for instance a butterfly or a bee, it can go the other way. The sea squirt starts out as a swimming tadpole with a brain. Once the tadpole finds a good-looking rock or coral bed to attach to, it becomes a plant. In the course of turning into a filter-feeding plant, they absorb their now unnecessary brains and spinal cords.

Sea squirt colony near Myanmar. Courtesy of Britannica.com

Putting aside that admittedly oddball exception, when nature makes a brain it usually keeps it around even when, as in the case of some worms, it could likely transition entirely to instinct with no noticeable effect. There is a funny high school science project involving 'training' earthworms to avoid certain stimuli to run a maze. Why would earthworms need to learn and remember? Who knows.

Nature may retain a brain in creatures not because it couldn't make every decision be instinct as it does with plants, but because it's easier to ramp up a tiny brain than to start a brain from scratch. A little brainpower is a useful thing to manage those situation-combinations that can confuse the best-planned instincts. When big changes occur, say an asteroid hits or some of the species gets trapped on an island, they can work that little pea brain for all it's worth to get though the day. Or the next fifteen generations.

Beefing up an already existing brain with 'more' genetic commands is a far quicker thing to do than starting one from nothing. When an environment stays within the range of cycling between hot and cold, wet and dry, sun and shade that is common over 200 or even 500 years, a species will have the right mix of instinct and brains to cope.

When the environment changes to something like it was tens of thousands of years ago, instinct reaches into the genetic stash for what worked the last time and turns it on. Maybe it was a higher tolerance for cold. Maybe it was an ability to get by with less sunlight. Maybe it was a darker red in the flower. Maybe it was flipping the 'more' switch for intelligence.

Intelligence is a stopgap measure. It's a desperate attempt to save a species when nature has no time to marshal instinct to cover the situation. Does it work? Does endowing a species with more intelligence at a crisis always get it over the hump? We know a handful of species surviving to today where we can say yes, it did. How many others went extinct when it wasn't enough? Unknown.

Hearing and Other Gifts

Transitional VS are in a very precarious position; until instinct takes over again, the hapless creature will have to think its way through things that other creatures in the environment are born knowing. It's not hard to spot examples of this in the wild world, even though we see only the proverbial snapshot of time.

Long ago when sea creatures transitioned to land, they managed it via an intelligence bump. While that was a very long time ago, on average land VS are smarter than always-sea VS even to this day.

Otters, porpoises, whales and sea lions are all smarter than their environment warrants because they experienced a second, much more recent transition from land to water.

These examples give us another clue about what nature likes to do. Nature could diminish intelligence when the species has had time to develop instinct, but sometimes it does not. If otters were smarter seven million years ago, we'll never know. What we do know is they have plenty of instinct now and yet are very intelligent for their size. Porpoises are still smart, with a language and a great capacity to learn. Yet their diet doesn't require any more intelligence than seen in sharks, which operate mainly on instinct.

Otters and porpoises are not 'going anywhere' with their intelligence. It's not getting stronger. We know this because they've each been in their environments for over seven million years and this is what they have.

Intelligence can be a stable thing in a species, like acuity in hearing.

Speaking of hearing, it sounds great to increase every species' hearing to the highest levels possible. Sure, it might improve survivability. But that's not how traits change. While there is variation in hearing ability between species, within a species it stays within a closer range. Some have good hearing, some not so good. But as long as they do as well as their grandparents, that's their hearing range.

Every individual of any species probably has moments when they wish their hearing was a little better. Nature finds that acceptable. With hearing, as with intelligence, nature tends to jack it up to only 95% of what the creature would like. Nature's Rule is not how much would be better, but if any less would be harmful. The environment only votes on done deeds. Any creature that isn't handicapped by their hearing will reproduce and could pass that same dimmed hearing to its offspring. Within each species hearing varies within a range, and if a species really needed it, if only those with hearing in the top 30% lived long enough to reproduce, hearing acuity would shoot up just like that. If the variation doesn't matter to reproductive success, it not only gets to continue, the range may even widen. Any month, any year that the environment starts voting that the hearing of 60% of the species is NOT good enough, that's the year the average takes a big leap up. The only other way it can happen is if a small family group that happens to be on the high side of the hearing range for the species gets segregated by geography and reproduces among themselves. Their average becomes the new average for their group, which in several hundred generations can become its own subspecies. Intelligence is like that too. If the environment votes 'not good enough' in a harsh way, it increases. Otherwise it stays the same, with only the in-species variation increasing.

Nature is very conservative when it comes to bestowing gifts, because Nature does not like excellent success. Nature loves the messy bird.

Deep Down

Simply swimming near a shore and eating fish doesn't require great intelligence; lots of dumb creatures, including other fish, do it. But when a monkey switches from a tree-canopy-life to a swimming life, what use is their instinct? They're going to deal with predators in a different way, need to identify what's safe to eat by trial and error and convey that to the kids.[27]

Respecting that there is no consensus in the field of anthropology on when we branched off, our ancestors became a shoreline living, fishing and scavenging species sometime between six million and two-and-a-half million years ago. We were the water monkey for a very long time.

---

[27] It's startling to parents that their two and three year olds will only eat what they've eaten before. Very often a parent can't get them to try something new without extreme bribery. Watching other children eat it–not other adults–is the best way to get them to eventually try a new food.

Just look into your own mind; if money were no object, most humans today would love riverfront, lakeview or oceanview living. We would love to fish or just lounge by the shore with all our friends and family nearby. Compelling feelings that are not created by our experience are instinct. Even if someone states they wouldn't like it, often a visit to the beach or lakeshore instantly changes their mind.

Must...put...feet...in...water...can't....resist.

There is nothing in our heads regarding staying up in trees for hours, of napping in the crooks of tree branches or going days without our feet touching the ground. I look at big trees. Is there an instant urge to go up, stay up there for hours? Watch the sunset from up there? Nope. Not feeling it. Dig deep . . .Sorry, nothing. It only seems uncomfortable. How about you–afraid of heights? Imagine a monkey who was afraid of heights. That's the opposite of being a monkey. We've been gone a long time.

What about big sky living? Well, that has some appeal. There's beauty in the big open ranges, of seeing so far into the distance. We 'enjoy' a view into the distance, an involuntary emotion that didn't get there from dwelling on tree-lined streams and ponds where seeing even four hundred feet away was rare. We like both.

The water life went on for a very long time. In that time horses and donkeys became separate species, zebras got their stripes, and we also underwent some profound changes in adapting to our new life. By 400,000 years ago we had language, used ropes and nets to catch food, and were using fire to cook, but we still lived in the same continent as our ape cousins. Around that time some of us trooped off the continent via the Middle East land connection. Another ice age glacier period descended, and the humanoids stuck in the southern parts of Europe and Russia became the Neanderthals. Trapped as they were in a chilling environment, they adapted to cold. They adapted to living in high altitudes. It cannot be a stretch to assume they did this with the help of clothes made of leather and possibly fibrous plants.

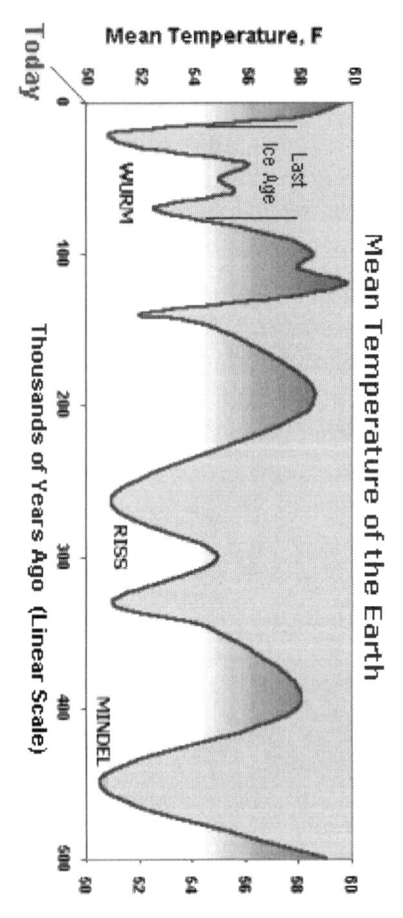

Ice ages are the dips, warm spells the peaks. Chart of the last 500K years. Experts disagree on starts and stops, but this is roughly the view held today.

It also cannot be a stretch to assume that none of their instinct from Africa really applied here. There were different plants, predators, conditions. They would have had to become smarter to survive.

All Neanderthals were descended from a rather small group, possibly around 100 individuals with some of them closely related. This is the most plausible way to explain their having a genetic variation only 1/3rd that of the people they left behind in Africa. Despite that shaky start, within 350,000 years their descendants had spread across all of Europe, the Middle East and Asia. This is a huge range to grow into in so brief a time. Most wildlife populations only replace themselves for tens of thousands of years. Neanderthals thrived; they had people galore, enough to split off family groups every second or third generation. They even populated islands many miles from the mainland, which could only be reached by boat.

Things were not so sweet back in Africa though. The same Riss ice age that started about 350K years ago and lasted over 100K years tied up all the water, reducing rainfall. Areas once lush jungle became . . . savannahs. Humans may have still lived along the water, but now that water didn't provide enough food. Their population plunged steeply. Scavenging and then outright hunting developed. Because none of their lakeside-living instincts served them well, they reached back genetically to something that worked in their last big transition from treetops to water: Greater intelligence. Like the Neanderthals, their environment changed severely, but they didn't travel to it; it simply changed around them.

Why . . . intelligence? Why not something more germane like leather skin and a coat of hair? Pondering for a moment, a less endowed creature than humans cannot be imagined. We were neither strong nor fast; we had thin skins and little hair. Our teeth were still fruit-eating teeth and our mouths hardly opened big enough to bite prey or predator; our jaws weren't strong enough to do more than deliver a bad pinch. Our sense of smell was weak and our hearing–well, we couldn't direct our ears to focus on a sound because each was on a side pointing in different directions. Human ears provided only the general direction of a noise, not the pinpoint location wolves and big cats could achieve. Our eyes were good–too good. We saw so many colors that perceiving movement was obscured. It took years of training for our young ones to be able to hunt and fish at all. Our arms had become much weaker over three million years; we couldn't hang our full weight from them for hours using just our grip force, nor could we walk on them. This was our inheritance: weak front limbs, not half as strong or as fast as any large predator, young that couldn't forage their own food for six to eight years, no claws, small mouth with dull teeth, bad sense of smell, lousy hearing. If that was your lot in life and the population was sinking fast, being smarter to come up with some tricks to survive was the best bet. Just being warmer, or growing claws, or some other measure wouldn't do anything for all the other feeble traits we had. No wonder people in Darwin's day scoffed at the notion that we were an animal. Nature would never put such a flimsy animal out in the wild.

Staying the Same, Changing

Scientists often talk about evolution as if it works like an apple grows. When ancient bones are found, that information is presented as a snapshot of a continuous process, like time-lapse photography would get of an apple growing from flower to fully ripe. Evolution doesn't work that way. Either some genetic change becomes useful, spreading around in a hundred generations like the FOXP2 gene, or the change is the reach-back kind that evinces like a blanket over the species in just a few generations.

Species stay the same, doing nothing but getting a broader range of genetic diversity, until a niche opens up or something happens. When a species leaps in to fill it, things happen fast.

Most species change quite a bit over thirty million years, but that isn't because anything bad happens. Mutations and genetic drift, i.e., species trying something out, occurs, and if they don't die young they get to try it again for another generation. The environment didn't 'need' horses, zebras and donkeys, causing them to be developed over the past five million years. Before them there were other grass-eating creatures, ones that either evolved to eat something else when these new guys came on the scene or simply learned to tolerate the company.

When humans came down from the trees and began swimming, they got an intelligence boost. It was enough to enable them do some things the apes never consistently did: make rope and nets; use sharp edges of rocks to cut; teach the young; and institute a system of older tribe members watching over the young. Perhaps using campfires to cook. After the initial intelligence bump, things stayed pretty level for 1.5 million years. Swimming and net-making improved their posture and their hand coordination. The look of their stone-chipped tools improved every 200,000 years or so.

To beat a dead horse, I'd like to try again to convey what a very long time 1.5 million years is. It means one does the same thing as one's great-great-great-great-great-grandfather, and he did the same as his great-great-great-great-great-grandfather, and then that goes on for another 7,142 times. Here were the early humans, just swimming and net-making and catching fish and watching over the kids. Not making any art, not singing, not telling stories, not reading the stars, not getting tattoos. Not evolving. If the climate hadn't gone to hell in a handbasket, they'd still be on those shores today, making nets, catching fish, watching the kids....

Then around the end of the Mindel glacier-period 420K years ago[28] or possibly the end of the glacial period before it,

---

[28] Glacial ends were marked by huge floods and many earthquakes and volcanos due to weight moving off land masses. Weight shifts allowed the release of tamped-down forces. In addition, rising seas and rivers changing course would have been horrific to those living on shores. Inland lakes might drain in a day. Rivers would start flowing in the other direction and become salty, killing everything.

the language gene took over. It swept over the entire population. If you couldn't talk, you just weren't going to make it with the ladies. Or vice versa. Planning and memory started to grow. A small group left Africa and headed north around 400K years ago, most likely moving along the Mediterranean shoreline.

These Neanderthals and Denisovans developed some physical and biological changes that helped them function better in cold climates, reduced sun and high altitudes. They used their intelligence bump to make boats, art, clothing, and musical instruments. Humans, on the other hand, retained their dark skin and evolved biological changes to help tolerate hot climates in full sun while standing on their feet most of the day. They used their intelligence bump to make hunting tools and coordinate hunting forays, as well as art, clothing and musical instruments.

After the next glacier period ended 60,000 years ago, humans, a new savannah-hunter with feet a bit more adapted to walking, headed north. Then we did what we do.

It's What We Do

Much of what is written about Neanderthals is slanted to justify or explain why we killed them. In photos we show them as ugly. We give them almost full body hair, black and messy–even though genetics is starting to hint that both Neanderthals and Denisovans had light reddish hair. We say they weren't a very good fit for their environment. They just weren't good at anything. They were stupid. They were too clumsy to make anything nice. Then in innocent tones scientists shrug, "we don't know what happened to them." Maybe we loved them to death, marrying them. Absorbing them like The Borg.

C'mon. We killed them. It's what we do. They didn't deserve it. We killed them anyway.

We kill the other. It doesn't even have to be anything major.

We killed millions of native Americans in premeditated biological warfare. Europeans actually took smallpox-laden blankets with them on a sea voyage just to make the natives in the new world die. Then we took their jewelry of gold and silver, their farms and their houses with the picket fences. They didn't deserve it.

In the WWII holocaust we killed five million Jews, over five million Soviet, Polish and Yugoslav citizens and POWs and 200,000 gypsies. They didn't deserve it.

Between 1975 and 1979 the Khmer Rouge in Cambodia killed or worked to death over four million people, approximately 25% of the population. They didn't deserve it.

In the two decades before 1908 King Leopold II of Belgium killed over eight million people in the Congo Free State via slave labor and disease. They didn't deserve it.

In 1994 in Rwanda the Hutu killed over 800,000 Tutsi in 100 days, plus many moderate Hutus who were opposed to the killing. They didn't deserve it. We killed them anyway.

The Irish Potato famine around 1850 happened due to exportation of food, not lack of it, killing a million Irish, about 12% of the population. The controlling English government could have ended it at any time, but chose not to. The Irish didn't deserve it.

The Armenian genocide began in 1915 and within ten years killed 1.5 million people. They didn't deserve it. We killed them anyway.

Russia (USSR) killed over twenty million of its residents between 1927 and 1953, many for refusing to hand over their property. They didn't deserve it.

In the 1937 invasion of China, the Japanese killed over ten million people, including Chinese, Indonesians, Koreans and Filipinos. They didn't deserve it. We killed them anyway.

It's what we do. We aren't 'better' or less vicious than we were 22,000 years ago. There are factions that are itching to do it again. They would kill millions with another religion, LBGT, immigrants. They say now they just want them to leave, but like in Rwanda, Germany and the USSR, that can turn on a dime and lead to door-to-door killing by people who were never very political before or don't even care to question why they are unleashed upon 'the other' in their midst.

Can good people stop it? No. In each case above, advisors, good people and religious leaders who opposed the killing were also killed. It's a death sentence to oppose what humans are hard-wired to do. There was never a genocide stopped by words of reason and exhortations to be kind. There may have been a few prevented by the timely deaths of the leaders orchestrating them.

We killed the Neanderthals and the Denisovans. That there were some pockets of happy co-existence was just as likely. Current theory–and it could change–is that human women partnered with Neanderthal males, and like matings between other related but diverged species, the female off-spring could reproduce well but the male offspring were likely to be infertile.[29]

All we need to do is look inside our souls and we know we killed the straight-haired, light-skinned, freckled, more robust to cold climates and high altitudes not-like-us people when we wanted their lakefront property. It's what we do.

That might have been where we stayed, a bright, far-traveling predator that wore jewelry, sang songs, drew on cave walls and made fat woman clay dolls. Officially, the Neanderthals and Denisovans were totally gone by 20,000

---

[29] The pairings could have been Neanderthal females and human males, too. Since children are raised by their mother, those children were more likely to suffer the same fate as their mother. The same male infertility situation would apply.

years ago. There is no evidence, but it's likely that pockets of them survived in inhospitable places, having no contact ever with humans, for several thousand more years. That wouldn't make a difference to the genetic story, though.

Next Ice Age Comes – And Goes

After humans left Africa 60K years ago and even got as far as Norway and Australia by 50K years ago, wouldn't you know it another ice age glaciation rolls in. Glacial starts happen slowly, so slowly that only the oldest humans would notice a slight difference in their lifetime. Glacial ends, however, involve every manner of disruption the surface of the Earth can conjure up. This last one peaked around 24,000 years ago.

During other glacial events, the clothingless African human stayed where it was warm. But this time was different; they had the Neanderthals who were already adapted to this kind of weather show them how to do it. Instead of high-tailing it back to Africa and warmer areas, they learned how to make clothes and those small stick shelters with a fireplace inside. They kept on spreading and even living not far from the edges of encroaching glaciers. What they learned from the Neanderthals was sufficient to survive. Interbreeding helped even more.

The dates of start and stop of all glaciers is mere guess-work, but of this last one we are certain: the world suffered through another glacial period end between 14,000 and 11,000 years ago.

By now the 'increase intelligence' switch in humans was a well-worn groove. In response to the deluge, the heavy rain, the earthquakes, the rising seas and the new rivers cutting a path without warning in mere days, nature evoked a stopgap measure one more time.

Until 12,000 years ago humans lived in small groups of no more than 60, had no permanent shelters and collected few belongings. They lived life akin to a wolf pack, interacting and cooperating within their group, friendly with a few neighboring groups to interbreed with them, but skeptical of strangers who spoke a different language. They loved their families and consecrated their dead with some measure of grief. They had a language, told stories, tended the fire and probably had food recipes. It could have stayed like that for another half million years. They didn't build things and they didn't collect things. They lived in the now. They remembered the past. They taught the young to hunt, to fish and to cook.

There's little evidence of ramping up intelligence between 50K and 13K years ago. There were slight improvements, indicative of a bit of training, passing along knowledge, but no brilliant breakthroughs that advanced the species. A thousand years is 40 generations. Between 50K and 13K no improvement happened swifter than 80 or 100 generations. If they were inventing anything or improving on methods and also could talk, you'd think these things would spread a few hundred miles in about three generations. After all, the people themselves were moving around, spreading far and wide. But not a single aspect of their lives improved with that speed. If someone from 15K was transported in time back to 30K, nothing would distinguish them as advanced. Maybe he or she now had a way to make fire from two stones instead of catching some from a wildfire, but that's a single generation discovery; either a person figures it out and shows his family, or doesn't. There isn't even any evidence that they wore shoes.

They made art. Yet for all that time the artwork varied more by the talent of the artist than how long ago it was made. Fifteen thousand years newer did not mean better pigments or techniques; these also varied by artist. Nothing but radio-carbon dating and other methods of analysis can tell us if this site is 15K old or 55K old. There's simply not enough difference in what they did or what they left behind.

After 10,000 BC, or 12K years ago, that changes. I don't even know where to begin. Simply everything started then. Houses. Doors. Money. Laws. Towns. Farming. Knitting. Shoes. Religion. Furniture. Indoor heat. Kings. Slavery. Pots.[30] Dishes. Bread. Beer. Wine. Farm animals. Not just once, but in several communities in Asia, the Middle East, Africa and Europe concurrently. Wow. Kind of puts going from radio to iPad in 100 years to shame.

That was just the first 3,000 years. Then came copper-working, writing, the wheel, accounting, woven cloth, taxes, far-ranging trade networks, plank boats with sails. In cities and towns there were people with trades–scribes, carpenters, cloth-makers, builders–who never grew or gathered any food, only bartered or sold their services or products for food.

Before 10,000 BC, humans, like animals, held their life's job to be about getting food. Look at any animal today; if they aren't actively seeking food or eating it, they are resting. Every human, like every single animal, sought food or rested three million years ago, 400K years ago, and 13K years ago. Except for the firetender, of course.

After 12K that changed; now it wasn't just firetenders. There were people who only made 'things' for other people and got food for it. That alone is the weirdest thing that happened within a head-spinning thirty generations 12,000 years ago. Someone could make furniture, or houses, or pottery all day long for other people who would 'pay' them in food.

After 4000 BC or 6,000 years ago, governments, brick building construction, palaces, irrigation and city-states

---

[30] Pottery for cooking possibly as old as 20,000 years is claimed to have been found in China and Japan. Whether this date sticks or not, both sites still say that farming and bread-making didn't start until 10,000 years ago.

became the way people organized their villages. Fast-food vendors set up shop in the streets selling breakfast to busy people on their way to work in towns around the Mediterranean and China. Stonehedge was built. Temples to the gods popped up wherever there was a tax base big enough to cover the cost. People got tattoos and had pet dogs and cats just for hugging.

Representation of the Tepe Gawra house on the Tigris river, built 5,000 BC, or 7K years ago. Was more or less continuously occupied for 2,000 years.

After 2500 BC or 4500 years ago, Egypt started to build pyramids and huge public buildings. They had rival nations and diplomacy and trade networks.

Once we turned the corner at 12K, no thousand years goes by without advancement in how humans live. Take a city person from 300 AD and plop them back to a city in 1000 BC and they would have amazing things to teach in every single aspect of life.

Stepping back, one can see how the ancient astronauts theory developed. A story about today's humans being a blend of alien and human DNA could get a decent foothold when the official story is that we were continually evolving for three million years and the first 2.6 million of them we really improved our stone knapping technique, but not much else. After that we could talk, yet it's obvious we didn't use talking to spread any knowledge. For another 400,000 years we made little sleeping hovels out of sticks with no architectural improvement at all.

Clay model of a poor city person's house in Egypt, 4,200 years ago. Found in burial crypt. Rooftop deck, private yard, double doors, windows. Around this time the rich had a bit larger city house, but then would also have a grand country house with decorative gardens and a pool in back.

If genetic change is slow, if nothing can be accomplished in ten or fifteen generations, then something magical must've happened 12,000 years ago. Sorry, no alien DNA appeared around that time. In fact, scientists are adamant that there is no physical or genetic difference between humans from 40K, 15K and 5K to account for the change. There is consensus, albeit mystified consensus, that humans had the same brain and were just as smart 40K years ago as they were 9K years ago. It's an elephant in the room that no one talks about because the answer could get weird: what the heck happened at 12K?

Not an intelligence bump. It was a regulatory control change. Why do I say this? Because if early humans came to

Australia and the Americas before 15,000 years ago, which it is pretty conclusive that they did, then they wouldn't have shared in the last 'bump.'

But there is nothing different about them. Their offspring today can go to school and become architects, lawyers and doctors. They aren't missing that last bump. Maybe it's not so simple. Maybe humans didn't travel to far-flung sites once-and-done, but came over repeatedly. For illustration purposes only, say some came 9,000 years ago, some 7,000 years ago, some 3,000 years ago, and the newer arrivals intermarried with the current population and spread around whatever was in that last bump. That didn't happen. For genetics to spread thousands of miles into hundreds of family groups and appear in everybody as a dominant trait, it would have been the most profound selective sweep ever seen in a species. But there's no evidence of that in the genetic record.

A simpler process explains it, one that we've seen in action in several species. In past chapters it was shown that a genetic feature now suppressed can be reactivated, or a 'more' command re-employed in very short timeframes, like flower color or animal size. There is preliminary evidence a regulatory control change can occur because the mom develops a strong feeling, much like mom turkeys realized humans can capture them yet not kill them, can be close yet do no harm, and *pass an attitude* to their offspring. Therefore, merely learning farming or herding or living in permanent housing could cause the change.

How can humans, purported to have the identical brain size, memory and speech ability 50,000 years ago that we have today, go 40,000 years just singing around the campfire, making, let's be honest, really clumsy jewelry, making the same tools over and over because they don't hang onto them, and then in the very next three thousand years, invent money and build houses with windows and furniture? Invent wood furniture and gardening and places of worship?

So what accounted for a jet-pack skyrocketing in so many aspects all at once?

Maybe it wasn't so many things all at once.

Maybe it was just one thing.

In 1995 there was a short-lived sci-fi series on TV called "Space: Above and Beyond." It had many interesting moments, but one of the premises was that the AI [Artificial Intelligence] human-looking things that they created as servants rebelled a few years back and almost won the war that ensued. The reason given for the source of their insubordination and independence of thought was that a guy inserted just one line of code into their programming: "Take a chance." The implication is that it made them more risk-taking and therefore inclined to improve their lot.

Not sure if the theoretical snippet of code, when talking about computer-controls, really extrapolates out into starting a war.

Despite my issues with that particular TV series back-story, it got me thinking: What if there was only time to insert one little snip of code during a crazy hazardous couple of generations for the human race? No time to change anything physically or biologically. Can't grow anything or remove anything. Just turn on a thought, or possibly beef up a mild thought so it is a primary thought. In three words or less. Just the proverbial snippet of code. Could this be done? Could nature do it?

I think so. But the important thing is, what was the genetic snippet?

If that snippet existed 10,000 years ago, it still exists now. Obviously whatever caused people to live in houses and join churches and go to school for 12 years and work the farms and pay taxes still exists today. Maybe this is where to look for the snippet; in things that our pre-12K selves would never do, but after that we all do.

So I started there with the Five Whys: Why do I pay taxes? So the community has money for big projects that no one person can afford to do. Why do I care if the community has money for big projects? Because things like roads and law enforcement and protection from invasion make my life safer. Why? Because there are forces out there that are scary and damaging: weather, bad guys, foreign powers. Why? Do you mean why do I think they're out there, or why do I believe my dinky contribution protects me? Because I have faith that all my neighbors and even those who hold power are united in opposition to those forces. We may differ in importance we place on different aspects, but we agree that larger entities provide protection and create community laws, buildings, roads. Why do I need those? Because without them I'd be afraid for my future. Why? Because all I really care about is my future. Most of my actions are aimed at taking care of my future.

The earliest cuneiform written documents, when translated, turn out to be land deeds, affirmations of ownership, and records of possessions. Laws seem to go back to the very beginning. Even if severity of the laws varied by rank, which they did, people depended upon laws to uphold ownership and decent behavior and then define a predictable punishment when violated. Writing was invented so we could say "Mine!" into the future. The earliest money tokens were I.O.U.s when the exchange of commodities was not even steven.

Start a Five Whys with farming, or buying insurance, or holding down a job, or having a car title, or belonging to a church, or start anywhere, with any aspect of our lives that is different than before 12K, and it boils down to one thing:

See the Future.

I must predict it; I must have a mental image of it that has me thriving or I work to fix it. I must make it secure. I must take care of today and tomorrow for myself, with tomorrow meaning next month and next year. In fact, for five years and even thirty years into the future.

Until 12,000 years ago, you could look back 1.5 million years or 30,000; we became more intelligent, had greater hand coordination, but cared about the same as any other animal for building something that would last two years. We cared about the same for having belongings: as much as we could carry easily, perhaps even with one hand, was the max.

A pet dog will have favorite toys and belongings, but does any dog devote time to making them? Are they particularly diligent in making sure they don't lose their toys? Imagine being much smarter than a dog but still with no great feelings about collecting items or making items. Something like a dolphin.

Imagine a group that can coordinate complex hunting efforts using their knowledge of the terrain to trap animals. Something like a wolf pack. That's what we were.

Then that changed. Even though the belongings are what shows up as artifacts, the belongings are just a symptom of the actual mental change. We didn't suddenly 'love' things. What we loved is counting on things to be there in the future. Owning something means having it both now and later. Possession, meaning it is ours now, is good enough for every other creature on Earth except us.

We need to know it can be ours, uncontestably, for as long as we like and even if we die we get to decide who gets it. Humans were almost indistinguishable from gorillas in their view of possessions 30,000 years ago or 1.5 million years ago. Then, within the span of about 1,000 years, we became different. All those years our ancestors chased after cows and horses and donkeys. Then, in a blink, they corralled them. This idea doesn't form slowly over 400 generations. It forms in one generation or simply doesn't. (A partial corral is nothing, and pointless to build. It has to go all the way around. Keeping animals in must be the intended purpose from the very start to justify the labor. There's no such thing as a one-winter corral; it's either built to last ten years or it won't last one year.)

If See the Future was not a switch that was flippable in whole tribes in less than 100 years, then there would be a gradual, spotty development over tens of thousands of years, as was the case with every single human advancement up until now: tools, hunting, drawing, rope-making, fire taming. Instead, what we see is a start in the fertile crescent (a term for the hospitable, green 'C' shaped area on the map swinging from northern Egypt to the Persian Gulf containing several large rivers for easy trade routes) and spreading through that region, which is a thousand miles long, faster than intermarrying and gene spreading could accomplish.

Even by assuming a 20-year human generation instead of 25, the existing population could not be genetically converted in that time. Nor was the area fully populated by a reproductively successful initial group of say 100 with the mutation, because then we'd see a narrowing of the genetic pool such as the Neanderthals had 400K years ago. The most profound selective sweep could not manage this kind of speed.

Houses and farming and animal husbandry were concurrent over a dinky period of time, spreading swiftly from this epicenter. Areas like the Russian steppes turned to mostly herding, not farming, while some areas had farming long before domesticated animals took hold. The urge to control the future food supply was identical; it simply adjusted for the realities of the flora and fauna in the area.

Within the human brain, in an eyeblink the future became more important than the present. How far will people go to ensure a good future? How much work will we do today to lock in safety, comfort and food for the winter, for next year, and even beyond our life, caring for those still living after we die?

As it turns out, pretty darn far. We'll work from sun up to sun down for months to get a good harvest. We'll work on projects taking weeks to hand them over to other people who will give us food and goods in exchange. We'll even work for decades at jobs we hate because it has a good pension. We'll work all year to have a one or two week vacation where we can do what we want. We quell every thought to stop going to work or not saving up to repair the roof by reminding ourselves it's good for our future, even though it is unpleasant today.

What else goes along with primacy of the future over today? It might lead to believing some people can see the future better than ourselves. Maybe not in all aspects, but some. Religious leaders arise.

That could be backwards; another possibility was that religious fervor grew the slow, genetic way over 400,000 years. This gave humans both a little long-term thinking and a little 'invisible larger power' thinking that, when both are cranked up double to quadruple, led to the human invention and easy acceptance of governments, militaries, kingships and other ruling bodies we concede to even though we may never meet them in our lives.

We now have faith that people living far away have power over us. Kings and lords fit our genetic model. We hope they will look after long-term goals while we focus on one-year goals; historically it has taken much malfeasance for us to oust them.

We anoint distant leaders with fame and wealth and extra rights. Living in a large house is almost mandatory. Look backwards in history or sideways to countries today; leaders build big houses and edifices for themselves.

When 'see the future' first appeared in humans, it was a response to a chaotic environment. It was a way to use that intelligence to store food, to improve food reliability, to build protective dwellings, and to devote time to making things that would improve comfort in the future. Around this time they even stopped tossing useful rocks at the end of the day; for the first time in the archaeological record we start to find rocks showing signs of continual use for generations. Prior to this we only find rocks reused, when found, a few dozen times over tens of thousands of years.

See the Future

Before 12K years ago, an area several miles in diameter was needed to support 10 to 40 people via hunting and foraging year around. As the group grew larger, nature couldn't replenish fast enough and certain plants and animals would become 100% eaten up within that radius. Groups had to split up before that happened. With short lives and high child mortality, this might be every three to eight generations. The split-off group may have to go very far away to find an uninhabited radius of the size needed. A way that groups could remain larger was by moving the whole group every six months. This allowed the old area to restore itself before they returned.

Any time small tribes moved, there was a risk of encountering hostile locals, who couldn't know this group was only passing through, not staying. Encountering angry bears and other predators was a danger every single moment. There was also a risk that when they returned to their old homestead next year, some other split-off group would be living there now.

Though travel was risky and life was full of the heartbreak of watching one's children set off never to be seen again, this was how humans, Neanderthals and Denisovans lived for over two million years.

Until 12K arrived.

After 12K, planting and tending to several patches of carrots, turnips, grapes, grain, apple trees, berry bushes and other plants produced enough food to feed 100 and no one had to walk farther than a mile. Towns developed simply because no one had to leave to eat. With enough food and daily interaction with more people, specialization began. Some people were good at making rope; others at carving wood.

Sociologists who have researched creativity note that inventiveness increases by the size of the nearby population. The more like-minded people that discuss and bounce ideas around, the more innovation occurs. All kinds of advancements were come-alongs when humans began living in groups of 300 instead of 30. Efficiencies of scale developed. This meant a person who liked doing a specific craft or process could do it enough for 100 people instead of for five. By getting really good at it, the end product looked much better. It made sense, for the first time, to spend eight hours building a workbench so the work could be done more comfortably and accurately.

Towns were safer. If four ruffians approached with theft in mind, it was not going to end well for a family group of forty people but twenty-two are under ten years old, three are elderly, two of the women are pregnant and four of the men were out hunting. When the same four come to a town of 300, with sixty adult men and women within hailing distance...the boys will have to behave themselves. Routine trade networks become more possible when a few strangers coming within a quarter mile do not put everyone into battle mode.

As the area converted to farmland expanded until fields butted up against each other, villages of 1,000 or more could be sustained. Towns that combined fishing, trading and farming grew to several thousand people–within just 2,000 years of the first farming that allowed a family group to enlarge beyond 60 people.

To put it in perspective, the bow and arrow was invented and then forgotten several times. The oldest bow found was made 60K years ago in Africa, but it took 50K years for the majority of humans to have that advancement, enabling the killing of medium-sized game over 80 feet away. The far more time-consuming and delayed payback of farming and building houses took less than 5,000 years to be adopted by humans in Africa, Europe, Asia and the Americas.

There is no evidence this change traveled via interbreeding passing along genes. Like the turkeys, humans changed in widespread fashion without losing genetic diversity.

No matter how long a gorilla lives in a house it will not wish to make one or leave it to the offspring. It will not learn to weave no matter how much he or she likes the blanket. Humans were roughly like that 13K years ago. Then, 12K years ago, we changed. Genetically, we now had a gut feeling that this would improve our future and we wanted to own it. Gut feeling, desire to own, this is the way instinct manifests: as a positive strong wish that we think is our choice.

Stepping back a minute from the headlong rush into cities and temples and long dresses on everybody, it's important to put the archaeological evidence in perspective.

The pieces we find aren't the best-looking, or the earliest, or even most representative of that time. They're the pieces that for some reason were handled differently than normal, were deliberately buried to hide them, were discarded because they broke, or were left on the battlefield with the dead. It's the house built where no one built again for a thousand years, covered with dust and leaves, then was built over.

We find the oddball, the one hidden or abandoned where no human laid eyes on it again the next year, or five years later, or even thirty years later. Or two thousand. Because if anyone had, they would have picked it up and taken it with them, to be lost to time.

We find remains like Ötzi, aka the Iceman, far in the alps and completely hidden by ice for thousands of years. Our best dating puts him living around 3200 BC, or 5,200 years ago, 5,300 by some accounts. By that time he carried quite a few items to make winter living easier, like a belt loop canister made out of tree bark that could hold live embers for hours, snowshoes, and a copper ax that would be handy for climbing icy slopes.

Scientists also found he had an intestinal parasite, and then in his pouch found a birch fungus which until recently was used as a medicine to eliminate that kind of parasite. Each part of his clothing was made from a different type of animal, apparently quite intentional. Genetic analysis of the leather shows the goat and sheepskin items were from animals who were ancestors to today's domesticated goats and sheep.

None of that–the copper tool, the domesticated animals, the medicine, the snowshoes with built-in boots–started only two or three generations prior. It took many generations to develop those items. The craftsmanship in several of the items, like the boots, indicate they were constructed by people who specialized in making them.

When we find the earliest instance of something, whether a copper tool, a sewn garment, or a house, you can be sure that was not when it began, but at least several generations earlier.

The Edifice

While specially prepared burials were happening for 50,000 years or more, with flowers and favorite items and a meal or two of their favorite foods included out of sorrow, after 12K there was another aspect to funerary items: things to take along to another world. Lots of things, not just a few objects the deceased liked while living. Bowls of food to last for a month. Pets and livestock. Some kind of backstory had developed which made tricking out a dead body with travel gear and supplies a sensible thing to do. The dead were no longer gone; they were heading forward to a new place.

Once See the Future was part of our psyche, then even the dead had a future. They not only had a future, they were marching into it. For the first time, human beings wanted to build things that would not only last their lifetime, but would last longer than a person could live. At first people built sturdier and more substantial homes, but within 1,000 years that segued into building huge things. At first these were burial mounds, man-made hills. Then they morphed into stacked stones, and then into temples. It never took more than 2,000 years to progress from mound-making to large buildings made with hauled and carved stone. In some cases it took far less time. The same progression occurred in Africa, the Middle East, Europe, South America, Central America, and then North America.

By the time of Stonehedge, concurrent with burial edifices in China, the Middle East, all around the Mediterranean Sea, Africa and Europe, it was so compelling to our ancestors to build a structure that would last 'forever' that it took the labors of several dozen people at least several weeks to construct. There is simply no connection between these and mere intelligence, yet the desire to build these spread in a somewhat radiating fashion to all the populations of the world.

This 'good idea' to build a huge edifice to memorialize someone important or to serve as a place of worship, or both, had everything to do with the future and nothing to do with devoting one's labors to bettering life today.

There was no survival-enhancing aspect to these buildings. For the initial few thousand years there were hundreds of stone edifices built all over the world which were not protective; they weren't places of defense.

What was so compelling? Perhaps that their work would say 'I was here' to the Future. For the first time humans not only wanted to see the future but wanted the future to 'see' them back. If everlasting wasn't the whole point they could have carved something beautiful out of wood which would be admired for five generations. When we look at the temples and burial mounds, we know they didn't make these because they liked someone in the past. They wanted to make a statement stretching into the future. "The future" became literally the most important thing. Today comes in second when it competes with the future.

After 12K, people collected food before it was needed and ensured harvests and meat enough to build this thing with the daylong labors of dozens of people. The edifice itself was proof that their seeing into the future was strong.

The farming and temple-building tendency came a little later to the Americas. It's almost as if the first wave of humans entering the New World didn't have it, but the later arrivals coming perhaps a handful of times between 8K and 3K years ago spread the idea. The oldest building in the new world is in Peru in South America, dated to 3500 BC, 5,500 years ago. By 100 BC mounds were being built in North America. By 1400, it had large edifices too. If one stuck color-coded pins on a map showing the age of burial mounds or buildings found in the Americas, it creates an amazing travel pattern radiating from Peru and 4K years later, arriving in the Southwest and then Central US, meeting up with the wave radiating from the East Coast.

Most of you are familiar with the walk-across the Bering Strait in Alaska theory of populating the Americas. While there's no doubt humans did that, genetics proves they also came from across the Pacific to South America and a bit later from Europe to the East Coast. The progress of See the Future tendencies flows in the opposite direction of the purported population flow from the Bering Strait to South America. The reason the Bering Strait theory was so stubbornly held onto by scientists even in the face of mounting evidence of human habitation twice as old is because . . .just what I'm pointing out happened at 12K. The change in human behavior after that date was so fall-off-the-cliff different that it created an anthropological dilemma. If the first residents of America came before that, it would mean they were still in 'shove stones, collect sticks, and drape with hides' mode for a place to sleep. So the population must have arrived after that because they were farmers, had houses, had money, wore clothes and had written languages.

The answer, of course, was the same as it was in the rest of the world. After 12K, when lost sailors or explorers or escaped fugitives or whatever landed in the Americas, the aspect that was so compelling and attractive as to cause the 12K changes in the Middle East also happened here.

Recent DNA testing on pre-1600 remains in North America turns up alleles that are only found in European, not Asian, populations. A theory being bounced around today is that along the edges of glaciers in the Atlantic Ocean there were seals, sea lions and other busy sea life going on, making sailing a boat along the edges and catching food along the way plausible. They could even park the boat at sunset and sleep on the ice. Humans may have come to the Americas often during a 30,000 year span. Only when the See the Future folks came over did there start to be easily findable signs of habitation.

Humans continued to thrive and spread. Farming, herding, fishing, food processing and food transporting was the full time occupation of more than 90% of all adult humans from shortly after 12K to around 1920. The other 10% could now be religious leaders, royalty, shopkeepers, tradesmen and teachers. After 1920, mechanization spread around the globe. Today in the US less than 5% are involved with food production; over the whole planet the average is less than 35%. The rest work at making goods and services they sell to other humans. Another portion are still in school, living poor for now so they will have a good future.

Stupid and Royally Stupid

When the future competes with the present, the future nearly always wins. In no other species is that true, and it definitely wasn't true of humans before 12K years ago. No other species has goals, saves up for years for something, or has eating come in second to getting 'things' one enjoys. Some people almost never allow 'now' to surpass the future. Men work long hours during their children's entire childhood under the conviction that they do it for the good of the children–meaning their future, not the actual kids today, who could be at this very moment riding dirtbikes atop fallen logs or smoking pot. Later, the kids don't like their mostly-absent Dad and he grumbles about ingratitude.

We all succumb to 'now' occasionally: the splurge, the impulse purchase, deciding one morning to take a day off work to go to the beach. If people do this too much or are too 'now' oriented, we think of them as untrustworthy. They stay in bed when it's time to go to work. They have high consumer debt with little to show for it. Impulsive people are frowned upon; those who enjoy 'now' too much are called hedonistic. Not looking into the future about each and every action taken is what bugs other people the most.

When someone robs a store that has cameras, or leaves something outside in the rain, or ignores a speeding ticket, or buys such an expensive car that they can't afford the insurance...and when asked, says "I didn't think about the future [the likely outcome]," we all have one thought: STUPID.

We don't think 'stupid' if someone is a car mechanic or a clerk at a drugstore and drives an 8-year-old car. We think stupid if their new car is repossessed because they couldn't afford that payment from day one. In other words, our definition of stupid is about an absence of contemplating the future most-likely outcome. If someone is forward-looking and living on modest means, never biting off more than they can chew, they're smart enough, they're fine. They aren't doing anything stupid.

When a person needs to be prompted to consider the likely outcome, or if in the aggregate their behavior won't lead to an expected future, it's then that we call a person royally stupid. We've all heard the line in movies and on TV: one person snarls at another, "Don't do anything stupid." It means, don't try to get revenge or steal your property back because it will backfire. We all understand the meaning as don't do anything to wreck your future. I can't think of a single instance of using the word stupid that doesn't mean making things worse by not considering the likely outcome.

IQ tests don't measure one's position on the scale of impulsiveness, meaning having a superior ability to momentarily suspend considering the future to engage in something enjoyable now. This is one reason IQ tests are inaccurate in predicting life success. It takes a lot of deferring gratification to be educationally and financially successful in life.

Some people, like ballet dancers, gymnasts and musicians make staggeringly huge investments at a very young age due to a consummate focus on the distant future.

Windfalls

People who collect a lot of money in a windfall fashion and six years later have blown it all are considered–let's be honest–to have something wrong with them in the head. They also think that themselves.

A windfall, such as from a lottery or inheritance, more often than not makes people less happy five years later than they would have been had their lives simply gone on without a big lump sum of cash. The reason is that it un-anchors them from their long-held view of the future. Goals of saving for retirement, of someday doing this or that, of working towards a promotion or a raise, of looking forward to and budgeting for the annual vacation, their next car or similar large purchase, all thoughts which grounded their every action and decision, are now out the window.

They replace the warm feeling of taking care of their future with the warm feeling of bestowing gifts on significant others, but it's a weak and fleeting substitute.[31] They are susceptible to the flattery and energy of 'business partners' who present a glow of purpose to replace their 'lost' future but in reality that situation is usually a scam. Windfalls are dangerous because they drastically change the future. Some people can handle it; most can't.

Which is why the advice to wait a few months before doing anything is so appropriate. It gives people time to weave a new future and nest the windfall into a normal life. All the 'stupid' behaviors repeatedly performed by windfall recipients are exactly what one would see if people lived like there was no tomorrow. With 'tomorrow' being defined as 'next month.'

They need to re-anchor themselves to a future where their actions make a difference. If they don't get past the feeling that their future is a lock and that it will be OK regardless of any actions today, they can in fact blow millions of dollars. No matter how much it is, it will be gone in less than six years.

---

[31] Sadly enough, the bestowing of expensive gifts on family members spreads the same ennui and lack of future-focus onto them, because they start assuming they can simply ask for money when they don't want to save up. It's almost impossible for a teen over 15 to get even a medium-sized windfall and then attend college and do well. A windfall inheritance ruins the future of young people, who don't have the perspective to know that an amount like $300K is nothing compared to good lifetime earnings.

Because humans are so focused on the future, tearing away our future literally takes our common sense away.

Money that's earned through work and planning can, but usually does not, unanchor the future. When the money is believed to be 'earned' it is treated as the intended future and handled normally. There are cases of millionaires via their own hard work and investments who won several million in the lottery, and within three years were hundreds of thousands in debt and declaring bankruptcy a few years later. Whether people are used to having a lot of money is immaterial; the unanchored future happens to them too.

Because this is a human reality that will play over and over, the world would be better off if lotteries awarded a maximum of two million dollars to winners; it's better to give thirty people two million rather than one person get sixty million. Two million is buying a nice house, giving a few family members new cars, and setting aside a nice amount for retirement. Sixty million is just cruel. Much of it will get into the hands of con artists and overcharging service providers, destroying the self-esteem and confidence of the 'winner' in the end. This is not 'maybe' or sometimes; when the amount is over $8 million, 100% of the time a huge percentage will disappear with nothing to show due to tales woven for the sole purpose of separating that 'winner' from their funds.

Statistics now show that 30% of people winning more than $3 million go bankrupt within five years. In a Florida study, 70% spent it all in less than five years. The normal rate of bankruptcy is 1% of households per year–but those people have often always been poor.

World Population

Twelve thousand years after becoming obsessed with providing for our future, we overpopulate the Earth. We have several cities of over ten million people. We have wiped out dozens of species in this time, most of them in the past 200 years. There are over 16,000 species on the endangered list, which includes virtually every predator species over 60 lbs.

Experts say if we continue using oil at this rate we'll run out of it in just over 50 years. Yet our goal is a low price per gallon.

You may not know it, but we are literally eating up the oceans. Fisherman have been ordered to stop catching certain fish or stop going to certain waters because species are so depleted they would cease to exist if fishing continued. Fishermen grumble, then they cheat. When you go to the store there may be six types of fish in the cooler, but when tests were performed by Consumer Reports on 190 samples from the greater New York area, less than 80% of those fish fillets were the labeled species. We are eating up even incorrectly-labeled fish from the rivers, lakes and oceans. Fishing fleet sizes are shrinking because there aren't enough fish to support the workers.

US Cities and states measure success by how much their population is growing. When a city's population reduces 2% over ten years, teams of people meet to come up with a solution. The only 'solution' the team should be pursuing is to break out the champagne and rejoice!

We were outraged at China's one-child policy, yet the whole world should be doing that. Sadly, China has now decided upon a two-child policy even though the former policy, not enforceable in rural areas, slowed but did not stem population growth. Two generations of 1.2 children per couple, or one natural and one adopted, would mean a world populated by 'wanted' children, with less neglect and poverty.

What would it mean to eradicate childhood poverty? Crime would shrink. Most violent offenders and thieves had miserable childhoods and half-hearted parenting. Yes, some nice people also suffered those, but statistically a decent upbringing doesn't lead to a life of crime. Simply requiring birth control to the age of 23, no babies before that, would vastly improve the lives of women, reducing the birthrate by 30% and inept childrearing by 80%. Birth control leads to wanted and loved children. I hope that is the point.

Even if the world population stopped growing today, froze in its tracks, the human population is still too large to sustain itself on Earth. There are quite a few renewable resources, like fish in the sea, but we are consuming them at a pace faster than nature can replenish.

The strangest aspect to population growth as the scourge of this planet is that we can't see it. Our minds are still in small-tribe mode, expecting that if our group gets too big some will just split off and go 'elsewhere.' Indeed, that was happening all the way up to the early 1900s. Then we ran out of new land masses to populate. Our genetic 'gut' feeling is that collecting resources and wealth is good for me and my children. We've now reached the point where hogging resources–big cars, heating huge houses, stuffing garages and storage sheds and rental units full of 'stuff' is actually stealing from your grandchildren. There is no more. There's a finite amount of steel, oil, aluminum, coal and copper on the planet. Use it up and your grandkids will have to pick through your 50-year-old garbage for tinfoil to remelt.

Pyramid Scheme

Currently, the only models for economic success and business success are ones depending upon more heads, more customers each year. I have seen many companies enjoy four or five years of sales increases, and then the next year sales go down to where they were three years ago–which was a record year and everyone got big bonuses–and it's a disaster, the company may fold. How can perfectly fine profits be starvation rations four years later? Something is wrong with this. Like a good household, exactly the same income should produce the same wealth. Possibly better, due to paying off equipment and developing better ways to do things.

We are smart. We ought to be able to figure out how to have a good economy while the population shrinks. All of our current economic models are pyramid schemes. They depend upon new blood coming in or the whole thing collapses. It's lazy beyond words to insist the only way to make a living is if there are twice as many people every thirty years.

Maybe our business models are pyramid schemes because those in charge want to make a killing, not make a living. More than once I've observed the published annual loss of a large company, in the millions of dollars, was actually less than the combined compensation packages of the top five guys. If they each took a salary of $500,000 and a company car that year, the company would show a profit. One could almost say the thumb on the scale that results in the eventual demise of every large company is the million-dollar salaries of the top guys. And demise they do. Very few companies are over 100 years old. They simply don't make it due to the pyramid scheme business model. Three years of making a shade less than last year puts them under. A functional model would be one that plans on using good years to pay down debt and save a little to weather the bad years.

People will never willingly accept a goal of reducing population by 30% in the next 50 years if every city's tax base, every state budget plus our social security system is based upon an ever-growing population. We have to make new models that work with a 3% reduction every five years. Banks have to accept those models, even demand them. We're smart. We can do this. It's not even difficult. Household budgets are very close to this design of prudent budgeting. A 5% reduction in income one year or a surprise $1200 repair doesn't mean we lose the house and live in the car. Businesses and governments must figure out a way to thrive with a decreasing customer and resident base.

Humans care about the future, but it's the way we 'care' for it that is actually killing it. The days of having multiple kids and walk-in closets and basements full of things are almost at the end of the line.

We turn to science fiction and fantasy. We humor ourselves, and many actually believe it, that colonizing other planets and sending ships to other solar systems dozens of light years away can be an answer. Never mind that if the pH varies from Earth's by 30% or there's a tad bit of methane in the atmosphere it means a full body suit for one's entire life. Besides, we have no clue how to travel at even 10% of the speed of light. We think scientists will come up with a way to colonize Mars to be self-sustaining. All nice thoughts, but we are out of time. We don't have the resources or technology to do it. One could say in three generations we'll be Star-Trek-capable, but we don't have three generations.

What we have are nuclear bombs. We have too many and there is not consensus about never using them. Eventually a few will be used. The magnitude of the ramifications are ungraspable by the average human. We think of them in terms of fireworks or movie scenes from WWII. The recollections of the two atomic bombs dropped on Japan at the end of WWII create a false complacency. After all, people live in those cities today. Tourists visit. People never stopped living there. The best of us imagine rattling our nuclear weapon capability as merely a scary threat, like two days of bombing raids taking place in ten minutes.

Popular Mechanics explains in detail how atomic bombs built after 1950 are from a dozen to over 3,000 times as strong as the ones dropped in 1945. Once a reader sees that, does every thinking human being drop everything to call their Congressperson and insist all these weapons be dismantled? No. The article is read, and then we move on to the next article.

If someone launches a bomb 3,000 times more powerful than the one dropped on Hiroshima, someone else will probably aim one the other way. This isn't just a bomb, this is Mount Everest becoming a lake. If one is aimed at Jacksonville FL, Miami is now on an island. A dead island.

Just being uninvolved will bring no comfort. Any weapon used will damage neighboring countries. The worst misunderstanding of the situation lies in not realizing the whole hemisphere will be affected by radiation and fallout. These weapons will cause tsunamis, earthquakes and possibly set off volcanos that otherwise may not have spewed forth for a thousand years. There is no do-over. There is no saying, oops, didn't mean to destroy your coastal city 1,000 miles away with a 70-foot tsunami. Can't say sorry, we didn't think so many volcanos would blow, shucks now none of us gets any sunlight for twelve years.

Nature didn't foresee any of this when she applied the stopgap measure multiple times on the same species. Who could know that last bump would destroy the majority of life on the planet within 15,000 years. All we were supposed to do is keep a stash of food, make a warm shelter, raise sheep and mind the weather. Instead we compulsively reproduced, cut down forests, fished the oceans empty and are on course to suck every bit of 500 million years of stored energy out of the Earth in 300 years. Then we'll launch some bombs that will poison the air for 100 years. As much as smart people may be horrified at that prospect, we created the bombs and are not eager to dismantle them. Eventually someone who is not a deep thinker will have the authority to activate one, and will. Can we keep the 'not deep thinkers' from leadership positions around the globe? No, we cannot. We either dismantle every last one or they will be our death.

Worse has happened and the Earth soldiered on. There were other times when a huge percentage of life on Earth ended. The Earth will recover from us too, even if we shoot off all the bombs. In ten million years it will be as if we were never here.

From 1960 to now we've grown from two billion to 7.5 billion people. There is no future for humans on Earth unless we can get back down to two billion. We must figure out economic models that allow households, towns, cities, states and nations to have comfortable lives and budgets as the population shrinks everywhere for 60 years. We must reduce endless acquisition of possessions and stabilize to a comfortable life of between 400-1200 square feet of living space per person. We must use our brains to switch as much as we can to renewable energy and material sources. That means we start making a hell of a lot of things out of bamboo instead of plastic.

We tripled our population in the last 60 years. We can live in any environment. We can talk to anyone, anywhere on Earth if we have WIFI. Even poor people have cars. We are so proud of being smart, with our telescopes and ski suits and indoor plumbing and libraries full of knowledge. We care about the future.

Can we figure out how to hold ourselves to a sustainable population that doesn't eat every fish out of the ocean, use up every drop of oil in less than 100 years, and kill off another 16,000 species by simply squeezing them out?

Can we see the real future that will be here in 50 years and stop it?

Say yes.

A New Look at Other Things

The previous chapter is a hard act to follow. Some would say, simply stop the book there. If you have more to say, write another book.

Yet I go on, because of the little-known fact that most non-fiction books of the past 50 years are seldom read to the very last page. Oddly, that resolves me to write this section. If readers are only going to read 2/3rds of the book, better they stop reading now than fifteen pages ago.

The next chapters take information that has been researched and published, takes facts that are known, and combines them in a way that sheds light on human nature, technology, and emotions. Not in that order.

The first part of this book has been about why we are smart, how we got that way, when did it happen, and the path to sustaining life into the future.

This last part may provide information that improves your life this year and helps you make better decisions. At the very least, I hope it proves to be thought provoking.

Weather Forecasting

In The Stopgap Measure, the weather in the decades around the receding of the last ice age was the key factor leading to a change in the way people thought. Seeing the future was about predicting the weather, a means to help us prepare and take precautions that would increase odds of survival. The weather, a.k.a. the environment by another word, was the cause of the change. But unlike other times when the environment caused a change, this time the change was a mental focus on the weather itself. A focus on seeing what's coming up. What's next. The change in us was a desire to take steps to predict and counteract negative environmental stuff coming in the future.

All the rest of the behavioral changes were an unforeseen side effect of working hard today to set in stores, make a strong roof and collect a good stack of wood in preparation for upcoming bad weather. They're what might be called a funny bounce, or in genetic terms, hitchhikers that came along with the desired adaptation.

Weather forecasting and mythic tales of control over the weather are part of all religions. Always have been, even in Christianity. If that is sacrilegious, may the clouds part and lightning bolts strike me down.

After 12K every society and tribe had a person or institution who foretold the weather. We became a species that believes people can predict it or influence it, either by special power or through sacrifices and obedience to God. Even today, we all watch 'the weather' on TV and listen to it on the radio. We have an app for it on our phones.

Yet the 5-day forecast is incorrect 70% of the time. For the past twenty years the Farmer's Almanac, which is published more than a year ahead, has been more accurate than 5-day forecasts in almost every region of the US.

In times when the weather is consistent, like summer in Wisconsin, if weather forecasters simply put 'sunny and high 80°s' as the 5-day forecast for the whole months of July and August, they would be correct about 50% of the time. Yet the guesses they place into the 5-day forecast end up only 30% correct. How can they be so colossally wrong and yet still get paid, still get airtime? Why aren't they trounced out for the fakers they are? They use their shaman-like radar and scientific words to lend themselves a cloak of competency. But they don't have to work the shaman aspect hard, because we hunger to believe. When they guess right about an impending storm or snow, less than 1/3rd of the time, we laud them. We just ignore all the wrong guesses because, well, we don't even know why. Weather forecasting must be real, it simply must.

What do I mean, they're 'wrong?' To be objective and use real numbers, say they are 'right' if the temp is within 3° and rainfall within three hours of the time they will say it starts and ends. Sunny or cloudy, simply that. For snow, within 3". Those tolerances are too tight, some might say. We can't hold them to that kind of accuracy.

Those are not tolerances I just made up. They are accepted norms for their own selected degree of accuracy. In other words, meteorology picked the tolerances, not me.

On a given day, say they state the high will be 83° F. Assume that the plausible range is 20° F., between 72-92° F. that day. Holding them to ±3% over the regional weather range, which would be 120° F. in Wisconsin, means their selected number could be anywhere within 30% of the plausible range and therefore be accurate.

What about the start of rain? There we top out at 24 hours; that means if they state an exact time, like 4:30, then

±3% is only 20 minutes either way. Calling it inaccurate only if beyond 3 hours is generous. In any other field except 'weather,' we expect experts to land upon it, to present a range that the actual occurrence has a 90% chance of landing within–not an under-30% chance. Expertise is defined as an ability to know the plausible range. Something weathermen cannot do more than four hours ahead.

If they intended to be honest and the best they could predict is mid-80°s or snow between 5 PM and 4 AM and between 2" and 7", they would say that. But they don't. They say the high will be 83, or it will start snowing at 6 PM and there will be four and a half inches of snow."

So are they right? And if not, why don't we care? We should. A lot of financial and social decisions are made based on their information, their certainty. Check it out: In any temperate zone where predicting weather would be a little bit important, like Chicago or Boston or Denver, collect the 5-day forecasts given daily for the past two years and compare them to what really happened five days later, using the tolerances mentioned above. The accuracy will be worse than using the monthly average.

No one sues them when schools and businesses close for a half inch of snow when they predict 10" or conversely, for the snows they fail to foresee that trap thousands of people in their cars.

News flash: weather prediction gets to 80% accuracy only four hours ahead. Which is about on par with what any 50-year old who has lived there his or her entire life could do.

I can't even watch the weather on TV anymore. They speak with such authority on Thursday that it will rain on Saturday that it sickens me. I know many people will be cancelling weddings and lawn parties, and odds are 50-50 that it will be sunny and beautiful on Saturday. What if they say it will be sunny and clear? Still 50-50. Because they simply do not know. They are talking out of their butt. That a guess is sometimes true is like a dice that sometimes comes up with a three. What we do know is that there are monthly averages for that time of year in that area of the country, and simply using those for the 5-day forecast produces greater accuracy than our current weather forecasters provide.

One year in Boston, about six times in two months the weathermen insisted big snow was coming and based on the word of those shamans, all the area schools were closed the night before. Mostly we got less than an inch by 9 AM the next day, perhaps 3", which is not serious snow in those parts. Conversely, the two biggest snowfalls that year were not predicted 50 hours prior. In one case schools scrambled to send kids home from school early because even at 6 AM that morning weather forecasters were predicting an inch or two, yet by noon there were several inches and now it was known it would continue for hours. Despite their 3-day prediction ability under 15% for *really* bad weather, no one holds them responsible for the harm they cause.

They are so bad at weather-predicting that at 10 PM their ability to predict tonight's low temperature within 3° is only 90%. This may be bend-over-backwards generous.

The role of weather predicting in modern society may be the weirdest thing an outsider would notice. Despite our demand for accuracy in every other endeavor, weathermen are allowed to blather wild guesses costing lives and money without even a harsh word. Would we contract a landscaper to mow once a week, and pay in full even though he did it only 40% of the time, then hire him again next year? Would we take our dry cleaning to a place with only an 80% chance of returning it to us? Would we pay a caterer in full if there was only a 50% chance they'd deliver any food? And two weeks later send them another check for another meal, even though nothing was delivered the last time?

No. It's preposterous. Yet we pin financial and social decisions on what the weatherman says and don't demand he cover the losses directly and solely caused by his services. He's held to be an expert, not to be contradicted or ignored, but is not held responsible for anything he commands.

It should go without saying: A person is LYING if they state on TV something is true but time proves it patently is not–and it was only a guess in the first place. That's a standard we hold in every single endeavor–except weather forecasting. He's given a platform from which we are told what to do, but nothing he says can be held against him in a court of law. He orders us to close businesses, stay home, do this, do that, and then, if what he says was the reason for those commands doesn't happen, somehow we don't blame him for making it up. Next week we fall for it again. And again. And again. And again. Thousands of times.

How do I know he made it up? Because it did not resemble reality. If he tells city governments to spend thousands of dollars in wages for snowplow guys who must be paid four hours' minimum for showing up and there's only 1/4" of snow, or none, too bad. His or her excuse? I do this all the time. And I changed my mind at 9 AM the next morning. You should have been watching TV so you could see when I changed my story. It's your fault.

We have this blind spot a mile wide around weather forecasting. It wouldn't be so bad if we really were better at it than 300 years ago. But it's about the same.

Sure, there are a few things; now we see hurricanes developing off the coast. We can identify cloud formations with tornado tendencies and give warnings. We're good at giving people four hours' notice of bad weather.

When fronts are moving with the prevailing westerlies, we can let people know when it hits their house there may be a 25° drop in an hour, or clouds are coming, or it could start raining. But that front is as likely to break up and never arrive as it is to stay together. After all their years of experience, they can't tell.

Whatever they say will happen more than 36 hours out is a maybe. We forget that something else happening is just as big a maybe. Whatever the weatherman says on Wednesday about Saturday's weather, a 3-day forecast, might happen. Whatever he says on Monday about Saturday's weather will not happen. I've lived the past twenty years using that as a rule of thumb, cheering when he showed rain, because using the 5-day forecast as what will NOT happen is a better prediction.

It's not just in the US; it's everywhere. When heading to Italy for ten days, starting from five days before I left, the 5-day out forecast said rain. It rained one day of my trip, 10% accuracy. Random guessing would work out better.

When 12K dawned we started looking for patterns, for signs that told us what was coming. We found them: dark clouds on the horizon, deeply-colored sunsets, wind changes, a smell. These gave us courage that other signs were there, we just haven't spotted them or interpreted them yet.

In fact, to this day almost every human being looks for signs, omens, premonitions, harbingers, prophecies and forebodings for almost every important endeavor.[32]

The place that our see-the-future genetic compulsion shows up most profoundly is in our never, ever calling the weatherman a fake shaman, a liar or a con man. Never, ever firing him over the damage he's done–even when it amounts to millions. Because what if he stops telling us the weather? What will we do? No one will turn on the news if we don't give the weather! We'll have to get someone else to tell us the weather. Do we really want someone WITHOUT Doppler radar telling us the weather? NOOOOOO!

---

[32] When I told my Mensan friends separately how, after deciding to move to Richmond, my first two home-buying attempts fell through just a day or two before closing, their immediate responses were "Don't you think that's a sign you shouldn't move there?" and "I'm surprised you're still going there. You should consider other places." They thought it portended badly. I was getting 'a message' that I was foolishly ignoring. There was no message, just two jerks. I moved to Richmond and love it here.

Besides, they all say the same thing, twenty weathermen on the radio, TV and newspapers in a region like Boston. Fire them all? Just for being incorrect? Just for being twice as inaccurate as giving the seasonal average for the region as the 5-day forecast? Just for not having the ethics and morals to restrict themselves to the timeframe where they can actually achieve 90% accuracy, merely the next four hours? Just for doing more harm than good? Why would we do that?

Blind spot. Mile wide.

Fatherhood

Going back 450 generations to 12,500 years ago, humans lived in small groups of related individuals. Couples paired up, and men were responsible for the offspring of their mate(s). While women can be certain the baby is theirs, all that engenders that same confidence in men is knowing they were present at the conception.

Today, men cannot be certain of that. Back before 12K years ago, men actually could be certain.

Going back a few million years, as human childhood lengthened, it became important for the father to invest time and energy supporting his offspring. Thus, very early in our development we began transitioning to pair-bonding.

Stepping back for a minute, pair-bonding in species develops only to secure certainty that the male is the genetic parent with concurrent assistance in rearing the young. A pair-bonded male will invest as heavily in the young as the female. In most species this means the pair almost never leave each other's sight. They are together all the time, which is how the male knows the offspring are his. But that wasn't a path humans could take because even at that early stage dimorphism developed in the tasks males and females performed.

Back then, just like now, men often stuck together most of the day. The men might go off for a few days at a time in hunting, fishing or foraging parties. How would a man know if someone else from some other tribe had lain with his wife while he was gone? Well, he wouldn't know. But it makes genetic sense to improve the odds that the pair-bonded male who expends a great deal of energy providing for and raising offspring has a compelling reason to feel he is doing this with his own biological children.

The great apes are not pair-bonders. We are evolving from a species that rather unfaithfully mates in opportunistic fashion to one that pair-bonds. We're almost but not quite there. It's likely our ancestors three million years ago consisted of females who mated with the guy who chased off all the other guys, and males who mated with all the women he could. The development of pair-bonding caused two changes in human physiology: one, love. Two, a way to ensure the male was the father.

This chapter does not talk about love.

Nature had a system in place to ensure human fathers would know. It was a pretty good system. It gave every father almost rock-solid certainty that the child born is his—or know with almost no doubt that it was not his. This system worked a million years ago, 400K years ago, and even 12K years ago.

A lot can change in four hundred generations, but only if there is some selection process going on. On this aspect, however, humans are oblivious, therefore no alteration has occurred.

What system, you wonder. As far as you know, there is no system for ensuring fatherhood other than a paternity test. That's because this old biological system doesn't work in modern society. It only works in isolated groups with three to perhaps ten women of reproductive age. What worked fine in small hunter-gatherer tribes does not work in cities and does not work in offices, factories, grocery stores or suburbs.

All this system is today is an annoyance to high school girls and to women who travel.

Here's how it used to work:

When women live together, their menstrual periods align. This is a biological reality. While there has never been any secret about it in medicine or folk wisdom, neither has it been assigned any particular value or significance. It falls into the realm of trivia.

My first major in college was Archaeology, and in a lifetime of reading about anthropology I don't recall this reproductive biological imperative ever playing a part in anyone's theory of early man. In actuality it's the only reason we could pair-bond, which enabled longer childhoods, and therefore enabled our higher intelligence.

Ponder it from a physical standpoint; something as profound as women's reproductive cycles having a never-fail synchronization policy cannot be just a random biological accident. If you read the earlier chapters on how the point of life is reproduction plus getting the offspring launched into the world and did not react with a feeling of outraged disagreement, then you have to slap-headedly admit that synchronized reproductive cycles in women living in small groups is IMPORTANT. It is a unique, universal feature of humans. Even the fact that these cycles roughly line up to the moon in the sky, just about the easiest way to track time, is significant. It is as if a greater force did that on purpose so people without calendars could tell when their next period would come.[33]

One of the uncommon features of humans is that our sex drives are not controlled by an estrus cycle; our sexual urges are always 'on.' Most mammals and birds have cycles ranging from once a year (deer rutting in the fall) to several times per year [dogs]. Sure, there can be dominance and territorial fighting at other times of the year, but these are only indirectly related to physical sex. Most species are very clear on the purpose for sex, and it's not for pleasure. For many

---

[33] Many cultures have a tradition of segregation during this 'unclean' time. It may have continued into modern times to give women some days off from unrelenting hard work, but it's likely to have begun long before clothes were invented, when blood would attract predators so it wasn't safe for them to be in the open.

species it's deadly business, as in the competitors may die. But when the switch is off, it's off.

The bonobo monkeys also have an always-on sex drive, but they have no need to pair-bond because their offspring have a shorter childhood. How does a species go from bonobo-style sex lives to pair-bonding? Well, via love, and then via a biological means to increase certainty that the male is the father of this pregnancy.

This system is the secret to why tribe-sized groups could pair-bond at all in a species whose sex drive is always on. While the physical desire of both sexes is always on, females have only a few days per month in which they can get pregnant. When menstrual periods align, then so do the fertile days. Due to the obvious nature of getting one's period in a small, outdoorsy group, both men and women knew the exact fertile days.[34]

When human women spend time together, an alpha or primary female pulls the cycle of the other females into alignment with hers via scent. None of the women do anything to make this happen and have no idea who is alpha. The means by which this happens initially is done by speeding up the onset by several days, but it can also be done by lengthening it a few days. After a pregnancy, a woman's period simply resumes in step with the group. No woman can choose a different time to be fertile. It is beyond her control.

The studies of this are pretty meager, but what constitutes an 'alpha' female as far as period timing goes may have little to do with social dominance. It goes without saying if the head woman is post-menopausal, that woman has no effect at all on their cycles.

---

[34] Anthropologists studying pockets of isolated cultures around the world after 1920 do not report 'everyone knowing' the fertile days of the month. This is partly due to getting responses to questions akin to 'the stork brings babies' when asked, and probably partly due to the interviewer not understanding or dismissing it as a fib that everyone knew which days a woman could get pregnant.

The useful function of this was that all the women in a group became fertile in the same few days every month. Because men require several hours between sex acts, this meant that one man would have difficulty fathering all the children in a tribe. A man who performs the act a handful of hours later will have an ejaculation relatively devoid of sperm. But never mind that. Concurrent periods means concurrent fertility, therefore concurrent pregnancies on a schedule that pretty much matched the moon. Even if the men traveled far and wide scavenging, it wasn't rocket science to plan to be home during the fertility week. A man who stayed in sight for the few days per month his wife was fertile could be certain all the children were his. All the men in the group were motivated to do the same.

This concurrent pregnancy tendency worked to everyone's benefit; women had nursing backup and the children had age mates to play with.

Since nursing moms might go 18 months between births, age group lumps were created in the tribe. Kids like age group lumps. They are hard-wired to be easier to teach in age group lumps. That's why two kids are such a handful and 25 kids are manageable by one person. That's why a single adult can get 25 kids to stick to a task for hours while trying to get one kid alone to do the same is a never-ending battle.

The reason humans evolved the way we did, why we can have schools where hundreds of kids obey two dozen adults for hours on end is because when women hang together several hours a day, their fertility cycles align.

If a five-day fertility window passed while the men were gone, no one should be pregnant. There should be a month of no pregnancies. The whole tribe would soon enough know if a pregnancy began while the men were out of town, because the fertility week was as common knowledge as observing the phase of the moon.

Since all the adult women knew the fertility week, even if they dallied with other men, it was possible to ensure all their offspring were their husband's.

What does it mean today? It means the menstruation cycle alignment to alpha females is still occurring, but with ever-changing amounts of women around it is harder to keep regular. It doesn't happen every time with every long bus ride or overnight stay in a hotel bed slept on by a different woman the night before, because the other women might not be more 'alpha.' Women who start a new job or travel often find their cycle suddenly shifts several days. But not always. Females with a strong alpha cycle will seldom to never experience any change under any circumstances. Women in the middle of the pack, not alpha and not the lowest, will sometimes have irregular periods.

Women who are very far from alpha will have irregular periods their whole life, because every day-long conference, day-long tour, new employee at work, overnight visit or hotel bed will pull their cycle one way or another.

For men, as a system for ensuring all the children are yours, it doesn't work. It's broken. For women, it means if you're going to travel, it doesn't matter what time of the month it is, pack some materials just in case. If you're starting a new job, wear something just in case on your first few days.

Grief

Grief. Why do we have it? What use is it?

Some mammals appear to have a form of grief. They are sad when another of their species or social group dies. They are sad when a human primary caregiver dies or disappears. But this sadness is not anywhere near the debilitating, handicapping, paralyzing thing to which humans are susceptible. In humans, grief can kill. Grief may last years, making a person unable to work to feed themselves. Making them a burden upon others. Grief makes people not take care of themselves, not bathe or brush their hair, resort to drink or drugs to distract from the pain. No other creature suffers a grief so profound that it would lead to their death if it were not for a social safety net.

Grief is humanity's doomsday machine. To someone looking at it for the first time, it is inexplicable. It takes fully functioning adults and makes them into basket cases when a child or spouse dies. Most people get over it in a few years, with struggle, but many do not. It pushes into their thoughts hourly, squeezing out their ability to fully engage in their life and use their intelligence to its best purpose.

Everyone will have a few years lost to grief. Some will suffer their whole lives from grief, one after another. Grief has a component of regret, and regret is thinking, 'I should have done something to prevent it." Actually having the ability to do something at that moment really plays no part; regret can be free-floating, as in, "If I was better to him/her, this wouldn't have happened.'

Grief may be related to the longer childhoods. When one child is born per birth and takes two years to nurse, there must be incentive to pay attention every minute to ensure survival. Grief is so destructive that a person would do anything to avoid it. It's so horrifying to glance at the chasm of grief one would feel at the death of a child that it's worth expending a great deal of effort to reduce the odds of that happening. Sell the house, divest of every earthly possession if that's what it takes, because the horror at living through the grief is so scary.

There was a dialog in a recent TV show that pretty much summed it up. The sad old man who'd lost a daughter several years ago said something like this to the police officer who was still trying to catch the killer, "Do you know why parents will say 'take me instead of my child' or put themselves in front of their child when a gunman approaches? Why they offer themselves instead of their child when they can? Not because they're noble. Not because they are selfless human beings. Not because they want to be a hero. Not because they calculate their child's life is worth more than theirs. They do it because *it's easier! It's less painful*!"

As a parent, I know he's exactly right. At that moment if I could pick grief and live, or simply go into that good night, I'd pick the second one. I'd willingly choose physical maiming over grief too.

Grief reduces conflict in social groups. People who are happy are feisty and opinionated; those in pain or grief are emotionally flat. They go along. They don't wander off, don't have initiative, don't get rowdy. They lack energy. They do the work they're told to do and don't make much conversation. Several cultures have noticed this side effect of pain, notably the Chinese and African cultures, so devised social customs that leave women in constant pain or discomfort their entire adult lives.[35]

The older a person gets, the more they hate the uselessness of grief. For instance, I want a good life for my son. The thought that grief, if I died in a car crash tomorrow,

---

[35] Chinese footbinding; statistics are that before 1910 more than 85% of adult women had bound feet. African cultures have a custom of female circumcision and sewing up labia of young girls, to be torn open on their wedding nights. In some areas of Africa there is a custom of severely burning a teenage girls breasts with repeated applications of a hot iron to prevent premarital sex.

would cause him to drop out of college, perhaps never go on to graduate school, develop a drinking problem, or any shade or variation of those typical grief behaviors, gets me angry. What the heck, I worked so hard to make him a fully-functioning, happy adult and now IT IS ME who throws a wrench into the works! Like taking a bullet for them, elderly parents wish they could absorb all the grief out of their children.

When elderly parents tell their children 'don't be sad for me,' they really mean it. They mean step off the grief train at the first opportunity, because I didn't give up a vacation in Greece to give you piano lessons, or move out of the city where I was within walking distance of a Walgreens to put you in a better school district only so you could drink yourself stupid in front of the TV for two years because you're SAD for ME. I did it so you could be a happy, successful adult, damn it! None of that sad baloney, get happy, buy a sailboat, join a choir, tend a garden.

The End of Potential Futures

Grief is the end of dreams. The end of the most optimistic hopes and plans involving that person.

Pick apart the actual thoughts composing grief and it's revealed to be a future loss. We're losing something that should have been in our future. So perhaps this debilitating sort of grief is an unintended side effect of Seeing the Future. Animals can have a sadness lasting a few weeks over an absence. When they don't know if a person is dead or just traveling, they can retain habits for a long time–going to the window when the person used to come home from work, behaving as if the person will show up again. But this isn't grief, it's habit.

Grief starts with a mental severing of all expectations and possibilities related to that person or situation over which one feels grief. As every string into the future is imagined, grief surges. We may not even have thought about a future interaction with them until it becomes impossible. It was just something in the multitude of possibilities. We could have gone here. We could have done that. We could have talked. We could have hugged. I could have improved my relationship with them.

How many times have you experienced the following: someone relays that a person you used to be close to, knew well or admired died a few years ago. You feel shocked, and then get teary-eyed. It's as if they died today. You experience grief even though you've had no contact with them for, say five years. Until that moment there was no grief; the grief launched only upon being told they are dead. What changed? Nothing. Had you gone another six months not knowing they were dead, your behavior would not have changed one bit. You may have tried to get in touch with them, but just as likely would not.

Grief is not simply knowing you will never see someone again. There are many people in our lives who we will never see again, but that pang of sadness isn't grief. It's more of an emptiness. It may linger for a few days and not be a happy thought whenever it comes up again for many years, but it will not hamper the ability to take care of oneself in the next few weeks.

You can pack up your stuff and move 1,000 miles away, grasping the concept that there are dozens of friends and relatives you may never see again, and yet feel only fully focused forward and energized by your adventure.

Instead, say the whole town suffered a gas attack while you were away the week before your move and those dozens of friends and relatives were dead. That would be devastatingly different. In both cases, say in the following 12 months you wouldn't have spoken to or had contact with any of them. In the first case you'd be fine, even cheerful, and in the second overcome with grief–yet the actual impact on your life is equivalent. Therefore, grief is a loss of a mental construct, not actually about not seeing someone or interacting with them. A close relative can go overseas or somewhere far away without contact for years and we may miss them, but feel no grief. Until someone tells us they're dead, then we weep for days.

Our grief may be too extreme because seeing the future plays such a large part in our psyche. We envision ourselves, strangely enough, going forward in life with all the people and pets we know. Each time that batch is diminished we feel grief. Usually grief is in proportion to the closeness or importance that person or pet held. But not always. When grief is held in or denied for someone significant, it can make us over-respond with grief to a death or loss in our outer orbit.

If grief is an aspect of our 'see the future' bent, then expecting the death can diminish grief, right? In actuality, this is true. People who know their dog breed lives on average about ten years can 'pull back' and mentally prepare for the end. They disentangle their mental image of their near future life with the presence of this pet. They take steps to 'fill the space' with new hobbies, travel, or another pet. They work hard to show appreciation and gratitude in those last days as an inoculation against regret. In those cases the debilitating type of grief may last only a week or two. After that it is manageable. A person may laugh easily at jokes again.

People whose family members have medical issues leading to death several weeks or months later have shorter and less severe grief, on average, than those whose significant other dies quickly and instantly. Some may say they merely started their grief earlier, while the person was still alive. Even taking that into account, the grief, statistically, is less debilitating when one has time to say goodbye.

The diminishment of grief, when able to say goodbye, is most profound with the people a person has known the longest: their parents. When the children and grandchildren truly grasp that all people die of old age, there was no keeping them forever, realize that it's a gift to parents to die before their children, and that only discomfort lies ahead for them, they may experience grief that soon diminishes after the death, washed out by compassion and the peace of knowing they spared their loved one more pain.

Length of grief has little to do with how much the deceased loved or provided for the mourner. It has to do with tickling the deceased out of their mental future and proceeding with the good future, even better than before, that the person would have wished upon them. In cases where it's a parent who wasn't particularly kindly towards their children, then proceeding to a happy life is just as important as it is for those whose parents adored them. There can be pleasure in having a better life than a parent predicted.

If grief is inevitable when we know people, how do people cope when entering situations where they know their companions might die?

This is exactly the situation for those serving in the military or as police officers. Entering service, most people start out by avoiding entangling these people into their mental futures. They may enjoy their company now and value them highly, but they avoid projecting a long future with them. If it happens, it happens. They keep the context of the relationship as short term. That may be a reason why meeting old military buddies is so enjoyable; it's a gift, not an expectation.

Large funerals for police officers reinforce the theory, not refute it. If each officer knows what he or she is doing mentally to prevent debilitating grief from destroying their career, then their worst fear is a 'who cares' attitude from their co-workers when they die. Both 'doing anything to prevent it' and 'huge inconvenience to everyone' then must become part of the institutional procedures instead of depending upon the grief response of all who knew him. The traditions of going after cop killers and pursuing the maximum sentence, plus all available police officers for 70 miles around attending the funeral are grief behaviors as company policy. Each person putting their lives at risk knows they can be trusted, and trust others, to treat co-worker's lives as importantly as if full-blown grief would occur.

If we do not get a handle on grief, it can ruin us. It's so easy to say to a grieving grown child, and so hard for them to grasp, that it's a doomsday machine, a poison pill, an emotion that will not bring back the dead but actually ruins the legacy of the deceased. If older people want successful and happy children, then they would no more wish grief on their children than would wish they lose a limb.

Government Funding of Grief Deferment

Extreme grief has always been with us, harmful in the single case. But now, in the past 50 years, a new ramification of the grief doomsday machine has materialized, and if left unchecked will destroy governments and civilization itself.

In the early days of medical research, from the 1850s to the 1950s, humans expected advances in medicine to improve lives, reduce pre-mature death, and make people happier. As effective medical treatments became widespread, that happened, but alongside the spread of vaccines and vitamin pills and routine surgery, another monster grew. Today it consumes 40% of every dollar the government spends on medical care.

What is it? Spending of huge amounts of money to stave off grief.

Consider how things often are: when a child dies their funeral is often attended by hundreds of people, but when a person over 85 dies, almost always there are just a handful beyond the close relatives. One would think the old person had a longer time to collect friends. That isn't what happens. Collectively we know that old people die and we disengage. It's extremely hard to get young people interested in visiting a person in a nursing home not because they're callous, but because there's no future in it. Why set themselves up to eat the grief poison pill? Instinctively they know that's not a good idea. We wouldn't strap a 20-lb. weight to our children and then say, 'try to get a track-and-field scholarship,' so why would we demand they develop fondness and then grief over a great-grandparent when their focus should be in setting up the foundation of their future lives?

Not everyone sees it that way. It only takes one person wishing to stave off their grief for just a few weeks to obligate the Federal Government to spend a million dollars or more.

Our current system of treating old age as if it were curable, until the moment it's not, is unsupportable. Strangely enough, it isn't even what the majority of people want when they approach death. When people are 50 they may plan and look forward to being 90, but when they're 86 with debilitating medical conditions or taking medications with miserable side effects, they see things differently. There's a point at which being miserable today isn't worth signing up for another of the same. If death will come, two additional weeks of pain is not a plus, it's a minus.

Profiteering From Grief

Part of the cause is the medical industry which receives the money the government pays. It has a financial interest in stretching out expensive medical care as long as possible. How many times have we all heard the same story from an anguished relative; they pop into the hospital room to find their elderly relative under-medicated or incorrectly medicated. The relative makes requests, protests, asks that it never happen again. The staff is bored with the hysterics. No worker is in a big hurry to fix the error. But five days later if that person, upon pleas from the elderly relative, states they are taking them home to die in peace, suddenly the medical staff is all over it, crowding in the room, threatening court orders, saying they will demand medical guardianship. Like they haven't already proved they are indifferent guardians.

Suppose the circumstance is that the person is 70 with a terminal illness or condition. There is no 'saving' them. All people die of old age. Another year or two in pain and discomfort is not a victory; it is cruel. Dying of old age means dying of simply whatever the old person comes down with in their weakened state.

Hospices have the right idea. Yet there are always adult children of very ill parents who demand every possible thing be done to prolong life. Even when it will not succeed. Many adult children acquire an insane tolerance, even eagerness, to cause physical pain to their parent; oddly, it's this memory of their complicity in this end-of-life torture that torments them the most later on. At the time they said they were doing it because they value life and value that old person. Later they realize they disregarded what the person wanted, and in fact what made lots of sense, just to inflict pointless torture whose only purpose was to stave off the onset of their own grief. Only in hindsight do they realize grief would have been profoundly diminished if they had honored the parent's wishes. It's even sadder when the parent worked hard their whole lives to leave a legacy to their children, wanted to pass on something of value beyond their death, only to watch it being squandered in torturing themselves in their last weeks.

Until about 70 years ago society or the government did not have to make rules about medical care. If one could afford it, one got it. If one could convince a doctor or hospital into providing it, one got it. The decision point we are facing today, how much should the government be required to spend on last-days care of the indigent elderly, is the first time in human history that we've ever come to this bridge.

When a 15 year old with no insurance comes in with appendicitis, no question it should be treated. When a 55-year-old has a car accident, absolutely. When someone who has bought their own insurance comes in with anything, at any age, their wishes should be followed. It is only when employing extraordinary methods to an elderly person who will go gently into a God-decided death if not fed by tubes, or hooked to IVs and breathing machines, or undergo surgery after which they will never recover the strength to walk again, that the question becomes: is this right? Is this a better use of government funds than any other thing?

In 2014 a facebook acquaintance related this 'funny' story–he called it that–which happened while his mother, over 80 and in the early stages of Alzheimer's, was hospitalized for weeks. Her hands were tied to the rails 24/7 to keep her from tearing out tubes; being unable to move led to bedsores and infections. This son came only once every day or two to find, always, she was under-medicated and in terrible, shuddering pain. She was thin and always cold; the staff provided only a thin sheet as cover. He was a party to this, in his assertion, because he loved her and wanted her to live. If I did that to my dog I'd go to jail.

His 'funny' story happened after her latest infection cleared and they removed the tubes from her throat. As he walked into the room she begged to leave like dozens of times before, but this time offered him one hundred dollars to take her out of here. He laughed at her.

When I wrote back that he should do it, stop inflicting pain and let her die, he accused me of cruel insensitivity. He wanted support during this difficult time, not have me siding with his mom. Several days later she died.

I hope you perceive the difference between actual medical care and whatever that end-of-life horror was.

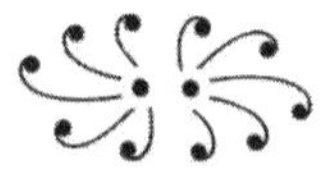

As things stand today, seniors over 85 will use 35% of their total lifetime medical expenditures in the next few years, however long or short it might be. In 2000, when adjusted for inflation, the amount needed per person for medical care between the ages of 0 and 48 was less than the amount spent after 85. With the ever-rising medical profiteering since then, it's probably twice as bad.

There's a difference between giving people help to get better and merely staving off grief. Most elderly people understand they've had a good, long life and want to call it a day. Remaining living by pulling huge resources away from kids, away from necessary infrastructure programs is not a good use of funds because *it will be unsuccessful*. The person will die whether they've taken $90,000 or $430,000 from Medicare.

In fact, most of the elderly DO wish to go quietly into that good night. My own Dad did. He always said he wanted to die in his sleep, just living life to the fullest to the very end and then not wake up. That didn't happen for him. He spent his last year in a nursing home after he lost a leg and his battle with cancer. He wore a 'do not resuscitate' wristband. He died only weeks before his savings ran out. Knowing his savings dwindled to the point where he had little to give his children and grandchildren was a sadness in his last weeks. He wanted his life, his terribly hard work, to mean more. He wanted to leave six months earlier.

Governments only recently began providing medical care. Before that, people either afforded medical care or they died. It's a new altruism for governments to step in so people don't die of easily-cured medical conditions, therefore our new dilemma, too. Now that our few decades of experience with it has shown that this altruism has taken a funny bounce, for the good of society and the country we must have this serious and extremely uncomfortable discussion.

We simply cannot continue to use government money to stave off grief.

Universal health care is good for society and good for individuals. But the intention never was that 70% of the funding goes to people over 80 who neither contribute nor enjoy life, simply painfully endure via contraptions we affix to them and drugs. People never recover from old age. The intention of health care is to get breadwinners back to work, to help children and young people lead full healthy lives, and alleviate pain near the end of life. Perhaps a sensible goal is to ensure that 75% of tax-supplied medical dollars go towards people under–pick an age, pick some rules–and limit those over that age to the remaining 25%.

Opponents say it amounts to 'playing God' and deciding who lives and who dies. We all die; that's God's plan. Foiling God is about denying his plan. It isn't *whether* we play God in extending people's lives beyond their natural course, it's *how much*. In a slow creep over 70 years, spending public tax dollars to play God has become mandatory. Worse, it has even become mandatory when the person who only sees bedridden misery ahead begs to be allowed to die. Can it be God's will to laugh at them, then bind their wrists tighter?

God

Stepping back for a moment, it must be acknowledged that no successful resolution regarding tax-funded medical care, medical standards or laws can ever happen without including God. Unless whatever is decided goes hand in hand with God's plan, it will not work. God designed people to die of old age, God's world has illnesses and injury, and God permitted all the medical discoveries that lead to restoring health in people who would otherwise have died if they'd been born 150 years ago. How much or how little a given person believes in God is immaterial. If the proposed plan seems opposed to what God would want, no one will like it.

One hundred years ago people decided that vaccines were in accordance with God's plan. God's world has disease, but God permitted the vaccine to be invented, so it met with His approval. The view is that God helps those who help themselves; passively waiting for mercy was not the only option, or even the preferred one. This line of thought leads to all kinds of nasty places, such as blaming the poor or afflicted for their health problems and believing if you are healthy that proves God loves you more, but most people are able to put a box around it. People don't deserve what they get, bad things happen to good people, but helping yourself to the best of your ability is never outside God's plan. It's always within the scope.

We must find a solution that is within the scope of God's plan that is merciful and respectful of the elderly person's wishes, never laughing at or dismissing their desperate pleas. Never denying care to those who need it to survive–but never prolonging discomfort just to eke out another few days, either.

A Life Well Lived

Years ago I read an article by a journalist blaming the Government for the death of his neighbor. He was pretty steamed. I would be too, if some red tape or negligence caused the death. I read on. As it turned out, the neighbor was 89 years old and a charming, sweet lady. She died during a hot spell in the summer. The journalist was outraged that the government did not buy her an air conditioner. She would be alive today, he said, if she had air conditioning.

Obviously, he was angry because he felt grief. He doesn't like feeling grief, so he's resentful and wants that lady to live forever. Just so he never suffers grief.

Another take on the same situation is that she lived a good long life. She had the huge privilege of dying in her own home, not a tube or wrist strap in sight. She was mobile and living life to the fullest to the very end. That's how I want to go. We all do. None of us want our last days to be peeing in bed on our bedsores and struggling for each breath, begging resentful staff for another blanket and they say they'll get one but an hour later no blanket yet. No one in their right mind thinks that's a better end-of-life than this neighbor lady, who never needed AC up to now, was able to accomplish for herself.

If I could go back in time, I'd suggest the journalist write the story as a perfect ending to a full life. The very type of ending he would wish for himself. Stop thinking the 'natural' ending is in a hospital hooked up to tubes. Try to imagine the grinding misery of it right now. When it's your turn, wouldn't you rather skip it?

It's great to help people through car accidents, wars, illnesses and infections, to patch them up and send them back into the ring. My own brother took a bad fall onto cement and died about six times in eight hours; each time the folks in the trauma center revived him. They did this even though they thought he would never walk or read a book again. Yet within three years he was entering 5K runs. He now manages his own business. It's not possible to determine how injuries will heal. It could have been the wrong bet; he could have been a ward of the state for the rest of his life instead of a taxpayer. It's vital to use medicine to the fullest extent. No man left behind. Yet old age is a different thing. I shouldn't have to explain the difference, yet I find myself in that position quite often.

Something that was common fifty years ago has left the American psyche: the idea of a life well lived. Now people only care about the amount of years racked up, not what one did with those years. The phrase 'led a good long life' is no longer in our vocabulary. When I say it, people wince. I mean it as a compliment, but it seems illegal to value a person for what they did with the years they had rather than just tallying up a large number. When I was young, a person who died at 60 could be praised for having lived a full life. Now one that dies at 72 has passed too soon, never mind how happy they were or how much of their bucket list they accomplished. It's as if one can't give credit for using the time wisely; they're all losers unless they make it to 90 or beyond.

People look at me strangely when I state that a life full of good works, loving children, and travel that ended at 65 is better than a lonely, empty life that ends at 85. For myself I want the life well lived, not simply a long one. And I absolutely don't want those last two weeks.

Nothing is sadder to a person who worked hard all their life and wished to leave something to their children to watch as all of it is pilfered away in a hospitalization that is only uncomfortable and lonely. At some point it isn't whether one has the money or not to pay for their own medical care, it's whether the children staving off grief is an adequate reason.

I've told my son a hundred times, money is for the living. Spending it in extraordinary care to ensure my last few weeks of life are particularly sucky and miserable, no thanks. I'll pass. I'd rather have it end in the few weeks before that when I was still mobile. Take the money and go to Key West, watch the double sunset. That will make me happy.

Last Chapter

"Our enemies provide us with a precious opportunity to practice patience and love. We should have gratitude toward them."

-- Tenzin Gyatso, The 14th Dalai Lama (b. 1935)

Three million years ago, 400K, 60K and 12K; I've mentioned these dates many times. These are merely representations of a date, a point when things happened fast around that time; someday it may be possible to pin it down better. It's almost certain it won't be these exact dates.

I admit I'm wrong about the dates, but not about the order in which things happened. By combining dig findings, nutrition, genetics, biology, psychology and physiology until all of them agree, it can't be dismissed by mere hostile insult, no matter how many degrees a person has.

The weak link in this, of course, is the dig findings. That's the aspect most likely to be bent by wishful thinking in the desire to produce something new. More often than not, new-found objects are declared to be a religious item simply because we can't figure out their use upon first glance. It's just as likely there were no religious items before 12K years ago. A dig can never 'prove' we were big game hunters a million years ago when our own tummies say nope, 150K tops, and more likely 80K; before that it was fish, fish and more fish. Digs can't prove we were hunting on savannahs with spears two million years ago when our own hands had webbing between fingers and toes until 300K to 50K years ago and it's still a dominant gene. Explain that, savannah holdouts.

This book couldn't have been written five years ago. The view is just materializing on how intelligence develops in all creatures and why we got the triple dose. I am cognizant of the irony on all kinds of levels of me typing the words on my computer, in my house with central air and city water, that the stopgap measure is a failure, an error, nature's biggest mistake.

Seeing the future is our gift and our curse. It's simply too painful to extrapolate what our future three generations from now will look like if the population doubles, and then doubles again, to 30 billion. We are too many.

Painful. Therefore, not a single world leader is laying it on the line for people. We hide in science fiction, in a fantasy that space travel is the solution. We browbeat our neighbors about recycling plastic when milk shouldn't even come in plastic bottles anymore, it should come in paper containers made out of bamboo. Most plastic things should start being made from bamboo.

Our government leaders, from mayors to Presidents, want population growth. Can they all be royally stupid? Maybe no one told them we're draining the seas of fish, have 50 years of fossil fuel left, and are squeezing 16,000 species out of habitat. Now you know.

In nature, the good is bad and the bad is good. Can we be smart enough to put aside instinct and reproduce at a .6 per person level (1 child per couple) for 90 years? Can we cease this insane race to consume every resource on Earth by backing off to four billion people, a sustainable amount for the long haul? If we can do that, we'll have ten generations to develop space travel and collect asteroids for their iron. If not, nature's lesson will be brains bad.

Bibliography

Photos

Page	Source
6	Lily patch http://www.gebs.net.au
8	Lily patch Wisegeek.com
19	https://commons.wikimedia.org/wiki/File:Fern_fossil.jpg; Picture taken by User:Vzb83
21	Image taken by Jim Peaco near Tower Junction, late August 1989, Nat. Park Services
24	https://commons.wikimedia.org/wiki/File:Platypus_in_Geelong.jpg Photo courtesy of TwoWings.
24	http://en.wikipedia.org/wiki/Thylacine Photo by E.J. Baker, 1904 from Smithsonian Institution archives.
25	Marsupial mole, Museums Victoria Collections https://collections.museumvictoria.com.au/specimens/121819
25	Golden mole http://www.biodiversityexplorer.org/mammals/afrosoricida/chrysochloris_asiatica.htm by Hamish Robertson
25	coastal mole http://mollen.info/
29	photo of moths on tree trunk https://commons.wikimedia.org Martinowksy from nl;
30	http://en.wikipedia.org/wiki/Peppered_moth_evolution
37	Rocket cannon net by: Richard S. Phillips, Graduate Student, Texas Tech University http://www.ranches.org/rio_grande_turkey_project.htm
38	Photo taken by Lynnette Hartwig
47	Frog on turtle Photo courtesy of Tanto Yensen http://earthables.com/frog-photographer/
50	Mitochondrial drawing, courtesy of OpenStax College CC BY-SA-3.0 via Wikimedia Commons
64	http://clinicalview.blogspot.com/2013/02/syndactyly-webbed-finger.html Photo by Mahavir Taylor, Feb. 2013.
75	Horse. Photo courtesy of Greater Ancestors World Museum. http://greaterancestors.com/horse-blinders/
76	Whale evolution Photo courtesy of Muizon C., Nature 413: 259-260, © 2001 Macmillian, www.nature.com

85	Bee Dance. Graphic courtesy of P. Kirk Visscher, University of CA, Riverside, in Encyclopedia of Insects (2nd Edition), 2009.
92	Hands. Photo courtesy of http://www.evcforum.net
94	Feet. Photo courtesy of https://sites.google.com /site/throwingevolution/
97	Otter. Photo courtesy of http://seapics.com/gallery/ Mammalia/Carnivora/Caniformia-Canoidea/Mustelidae-Mustelids/Lutrinae/North-American-river-otter-search.html
99	Proboscis monkey By Charles J Sharp, from Sharp Photography, sharpphotography, CC BY-SA 4.0, https://commons.wikimedia.org
100	https://commons.wikimedia.org/wiki/File:Snow_Monkeys ,_Nagano,_Japan.JPG Photo taken by Ybleib
101	Hamadryas baboon photo http://www.yourdictionary-.com/baboon Photo courtesy of Alamy/Arco Images.
102	Olive Baboon face https://commons.wikimedia.org-/w/index.php?curid=29419665 By SajjadF
103	Sean Connery, public domain.
104	Jack Black. Photo courtesy of Kevin Winter/Getty Images.
105	Photo, public domain
122	Terra Amata sketch By José-Manuel Benito https:// commons.wikimedia.org/w/index.php?curid=548117
136	Sea Squirt colony. Courtesy of Britannica.com
141	Ice Age chart http://www.aerobiologicalengineering.com /wxk116/StoneAge/Habitats/
151	Round house http://www.artefacts-berlin.de/das-round-house-in-tepe-gaura/?l=eng
152	Egyptian house http://antigoegito.org/as-casas-egipcias/

References

Allard, C.,, R. Mathieu, G. de Lamirande, and A. Cantero http://cancerres.aacrjournals.org/content/canres/12/6/407.full.pdf Mitochondrial Population in Mammalian Cells by Nov. 27, 1951

Alemayehu, B and Warner, K. E. https://www.ncbi.nlm.nih.gov/pmc/articles/PMC1361028/

BAINES, John; MALIK, Jaromir. Cultural Atlas of Ancient Egypt. London: Andromeda Oxford Limited, 2008

Berna, F., Wonderwerk Cave https://www.youtube.com/watch?v=uRj9H96a89E

Best, B (2003). "Ostrich Facts | access." The New Zealand Ostrich Association

Brown, T.A., 2010. Stranger from Siberia. Nature 464: 838-839.

Coop, G., Bullaughey, K. Luca, F., Przeworski, M., 2008. The timing of selection at the human FOXP2 Gene. Molecular Biology and Evolution 25(7): 1257-1259.

Despammed, O., http://en.wikipedia.org/wiki/Peppered_moth_evolution

"Evolution of Federal Wildland Fire Management Policy" (PDF). Review and Update of the 1995 Federal Wildland Fire Management Policy January 2001.

Ferentinos, G., Gkioni, M., Geraga, M., Patatheodorou, G., http://www.sciencedirect.com/science/article/pii/S0305440312000441

Journal of Archaeological Science Vol 39, Issue 7, July 2012, Pp 2167–2176

Flatt, A., https://www.ncbi.nlm.nih.gov/pmc/articles/PMC1200697/

Franke, May Ann (2000). "Changes in the Landscape" (pdf). Yellowstone in the Afterglow. National Park Service. Retrieved 2007-08-03.

Franke, Mary Ann (2000). "The Summer of 1988" (pdf). Yellowstone in the Afterglow. National Park Service. Retrieved 2007-07-17.

Garcia, J. M., http://www.smithsonianmag.com/history/german-pows-on-the-american-homefront-141009996/

Green, R. E., Krause, J., Ptak, S.E., Briggs, A.W., Ronan, M.T., Simons, J.F., Du, L., Egholm, M., Rothberg J.M., Paunovic, M., Pääbo, S.,. 2006. Analysis of one million base pairs of Neanderthal DNA. Nature 444: 330-336.

Hawks, J., http://www.slate.com/articles/health_and_science/science/2009/02/how_strong_is_a_chimpanzee.html

http://humanorigins.si.edu/evidence/genetics/ancient-dna-and-neanderthals/sequencing-neanderthal-dna

http://www.octopusworlds.com/octopus-reproduction/
http://www.pbs.org/wgbh/nova/neanderthals/mtdna.html
https://www.fws.gov/mountain-prairie/species/mammals/grizzly/grizz_foods.pdf
Krause, J., Lalueza-Fox, C., Orlando, L., Enard, W., Green, R.E., Burbano, H.A., Hublin, J.-J., Hänni, C., Fortea, J., de la Rasilla, M., Bertranpetit, J., Rosas, A., Pääbo, S., 2007. The derived FOXP2 variant of modern humans was shared with Neandertals. Current Biology17: 1908-1912.
O'Neil, D., http://anthro.palomar.edu/homo/homo_4.htm
Overbye, D., http://www.nytimes.com/2011/07/28/science/28life.html
Pääbo, S., https://www.youtube.com/watch?v=uyItUq1uy7c International Harry Steenbock Lecture 12-12-16
Patterson, N., https://www.youtube.com/watch?v=WddkiPx7QR0 Science for the Public's Working Science Series, July 08 2016. Nick Patterson, Ph.D. Senior Computational Biologist at the Broad Institute
Pontikos, D., http://dienekes.blogspot.com/2008/10/60000-year-old-y-chromosome-haplogroup.html
Pottery, Ancient, http://www.visual-arts-cork.com/pottery.htm
Rosas, A., Pääbo, S., 2009. Targeted retrieval and analysis of five Neandertal mtDNA genomes. Science 325: 318-321
Rothman, M.S. 2002: Tepe Gawra: The Evolution of a small, prehistoric center in northern iraq, University Museum Monograph 112.
SHAW, Ian. The Oxford Illustrated History of Ancient Egypt. Oxford: Oxford University Press, 2000
Smithsonian Institution, http://humanorigins.si.edu/human-characteristics/tools-food
Spain, D., http://www.pbs.org/fmc/timeline/dmortality.htm.
Stewart, D. (2006-08-01). "A Bird Like No Other." National Wildlife. National Wildlife Federation.
Templeton, A., 2005. Haplotype trees and modern human origins. Yearbook of Physical Anthropology 48: 33-59.
Tobler, A.J. 1950: Excavations at Tepe Gawra, Bd. II, Philadelphia.
Turner, Monica; Romme, William H; Tinker, Daniel B (2003). "Surprises and lessons from the 1988 Yellowstone fires" (pdf). Frontiers in Ecology and the Environment. 1 (7): 351–358. doi:10.1890/1540-9295(2003)001[0351:SALFTY]2.0.CO;2. Retrieved 2007-08-03.
US Fish and Wildlife Services, https://www.fws.gov/mountain-prairie/species/mammals/grizzly/grizz_foods.pdf
Zorich, Z., "Timelines" Archaeology Magazine Vol 69 No. 4, July/Aug 2016, pages 33-35

Index

Made in the USA
Middletown, DE
19 September 2023